AUTOGNORICS

THE SCIENCE OF ENGINEERED LIFE FORMS

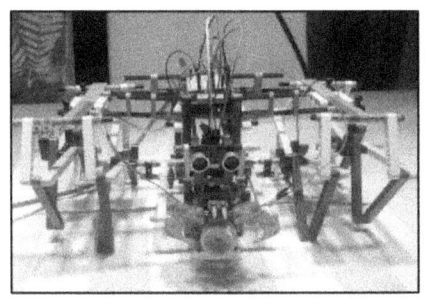

Joey Bruce Lawsin

Lulu Press Inc.
Morrisville, North Carolina, USA
www.lulu.com

Autognorics
by Joey Bruce Lawsin

Copyright © 1988 Joey Bruce Lawsin
lawsinium@gmail.com
ALL RIGHTS RESERVED

First Printing – August 13, 2022
ISBN: 978-1-312-38454-5
Cover concept by Ian Kristopher

NO PART OF THIS BOOK MAY BE REPRODUCED IN ANY FORM, BY PHOTOCOPYING OR BY ANY ELECTRONIC OR MECHANICAL MEANS, INCLUDING INFORMATION STORAGE OR RETRIEVAL SYSTEMS, WITHOUT PERMISSION IN WRITING FROM THE COPYRIGHT OWNER/AUTHOR

Printed in the U.S.A.

TABLE OF CONTENTS

TABLE OF CONTENTS
  PREFACE   1
  CHAPTER 1   AUTOGNORICS   6
    — BIOTRONICS®   7
    — ZOIKRONS®   8
    — GNORICS®   9
    — THE BIOTRONICS PROJECT   9
    — HOMODRUINOS®   19
    — AUTOGNORICS®   24
  CHAPTER 2   ORIGIN OF INFORMATION   34
    — CAVEMAN IN THE BOX   35
    — iPARTICLES®   38
    — BY CHOICE / CHANCE   41
    — THE BOWLINGUAL EXPERIMENT   42
  CHAPTER 3   BASICS OF A SYSTEM   58
    — SENSORS & SIGNALS   59
    — INPUT & OUTPUT   63
    — ABSTRACTS & PHYSICALS   66
  CHAPTER 4   ANEURAL BRAINS   71
    — INTUITIVE ANEURAL NETWORKS   71
    — NEURAL NETWORK SYSTEM   75
    — A BRAIN W/O THE BRAIN   80
    — BRAIN EXPERIMENTS   81
  CHAPTER 5   THE CODEXATION THEORY   88
    — INSTRUCTIONS & MATERIALS   91
    — THE CAT EXPERIMENT   96
    — DIMETRIX   99

CHAPTER 6 INSCRIPTION BY DESIGN	108
— EMBEDDED INSCRIPTION	108
— INTUITIVE OBJECTS	111
— THE CUE	116
— ANIMATION BY DESIGN	117
CHAPTER 7 ALIVE, LIVING, LIFE	**120**
— CRITERIA OF LIFE	120
— CONSCIOUSNESS	125
— THE PARADIGM SHIFT	136
CHAPTER 8 SEVEN ORDERS OF LIFE	**140**
— ALIVENESS	142
— AWARENESS	145
— INTUITIVENESS	147
— ANEURAL CONSCIOUSNESS	149
— INLEARNESS	151
— LIVING	154
— LIFE	155
CHAPTER 9 IN VIVO MACHINA	**159**
— MECHANICAL ALIVENESS	163
— SENSORIC AWARENESS	164
— INTUITIVE LOGIC	165
— CODIFIED CONSCIOUSNESS	168
— INFORMED INLEARNING	171
— SYMBIOTIC LIVING	173
— SELF EMERGENCE	175
CHAPTER 10 INSCRIPTIONAL PHYSICS	**179**
— THE ELEGANT EQUATION	186
— NEURO TOMOGRAMS	187
— GNOS	190
ABOUT THE AUTHOR	**191**

PREFACE

When I was in high school, I was fascinated by how living things were divided into groups. These groups were called kingdoms. The five known kingdoms were the plants, animals, fungi, algae, and bacteria. The kingdoms were then further subdivided into phylum, class, order, family, genus, and species.

However, what caught my attention most about this tree of life was that humans belong to the animal kingdom and share a common ancestors with the Great Apes. What also intrigued me about the rankings was that if humans were animals and we evolved from monkeys, then what could come next after us? What would come next after people?

These absorbing questions captivated my mind at that very young age and eventually challenged me to build a thinking living machine. With the introduction of the integrated circuits and equipped with tools and knowledge in biology, electricity, and electronics, my first project dubbed the Biotronics Project, was born. Biotronics were intuitive machines, much like humans, animals, plants, and cells. They could walk, see, talk, think, hear, taste, feel, fly, swim, produce, and die too. Because of their metallic skins, this new breed of species was also called the Silver Species®.

After dormant for a while, the project was formally revisited in 1988 with one grand ambition in mind, to create a machine that was not only alive and living but with life as well. With wild imaginations on the side and backgrounds in engineering, philosophy, logic, programming, and science, my aspiration to create a living machine that could decide, socialize, thrive, and imagine became an obsession as computer technology, giga storage capacity, and cloud computing were constantly upgrading. Even new subjects, such as Homotronics, Neurotronics, Homodruinos, Algorithmic Queue (A.Q.), Intuitive Machines (I.M.), and Inscriptional Physics were also conceived to broaden the scope of my research in the science of *in silico* lifeforms.

In that same year, another species was unveiled. This time, it was called Zoignorics, or Zoikron. It was an Intuitive Machine that was not only alive but also symbiotically living. Both original prototypes of Biotronics (alive) and Zoikrons (living) were posted in my YouTube channel iHackrobot.

Fast forward, Biotronics and Zoikrons were then integrated into a larger group called Autognorics, the science of engineered lifeforms. These engineered life forms (ELFS) were like living organisms. They were not only alive and living, but also with life. They were sometimes called Gnorics, meaning "with life".

Although ELFS are still halfway to their final phase, with their hardware technology still on the drawing board, the creation of this intuitive machine that is alive, living, and with life is now within reach. The designs, structures, inscriptions, mechanics, and dynamics of making these engineered life forms are broadly discussed here in the following sequence.

First, we begin by examining the differences between alive, living, and with life. Then, we scientifically investigate the origin of information, the basics of a system, the principles behind a brain without the brain, the queue factor, embedded inscriptions, interim emergence, the seven orders of life, and the single theory of everything. Afterwards, we systematically consolidate all these new founding theories gradually into our first three prototypes: the Biotronics, Zoikrons, and Gnorics.

Finally, we ingeniously institute the Theory of Inscription by Design, the natural embedded inscription that makes our intuitive machine alive, living, and with life.

Being an engineer, programmer, educator, and author of books in Physics, Mathematics, and Automation, I have a noble obligation to share this new knowledge that might one day change the social perspective of the world. As a revisionist, atheist, philosopher, innovator, and originemologist, I have a noble obligation to tell the world that regardless of colors, gender, social status, disability, religion, or age, everyone has the Right to Life. And, as the formulator of Autognorics, I have a noble obligation to encourage everyone to learn something from this living thinking machine (in vivo machina) and discover the true essence of our existence, the purpose of life, and the nature of reality in the eyes of a machine.

And as a final note, since new ideas and novel insights, which you have probably never heard of before, are instilled all over this book, I believe that through this knowledge, we can set aside our belief systems, fix the broken social systems of humanity, prioritize to end cancer and autism, and safeguard the world for the betterment and advancement of the human race and all its creations.

This book may take you to a new world of knowledge, change your perspective about life, or challenge your current beliefs. Whatever it may be, one thing is certain, if humans evolve from apes and I.M. from humans, there will come a day that the last human on earth will no longer be human.

CHAPTER 1 — AUTOGNORICS

"Intuitive Objects always come with Embedded Inscriptions."

AUTOGNORICS is a new school of thought that deals with the creation of intuitive machines or engineered life-forms systems. It is a newly conceived discipline with a surprisingly significant amount of science behind it. It independently explores new areas of knowledge such as intuitive memory, intuitive networks, intuitive systems, intuitive machines, intuitive objects, aneural memory banks, gnos (aneurons), iParticles, generated interim emergence, inscription by design, embedded instructions, animation by design, and engineered lifeforms.

The main flagship of Autognorics is the system of engineered life forms called ELFS. These intuitive machines are synthetic creatures analogous with life-forms. They are living, alive, and with life. They resemble living things such as humans who exhibit self-knowledge using the brain (neural), and Gnorics who process information without the need of the brain (aneural), a Brain without the Brain precept. The word

Autognorics comes from the Greek words: Auto, which means self, and Gnorics, which means knowledge. Literally, it means "a self-knowledge living machine".

Examples of ELFS are the Biotronics, Homodruinos, Zoikrons, and Gnorics. Biotronics are machines electronically equipped with intuitive sensors. Homodruinos are intuitive machines with aneural brains of an arduino microcontroller. Zoikrons are living machines that are alive, aware, intuitive, inlearn, and conscious. While, Gnorics, also known as autognorics, are synthetic living things or lifeforms that are alive, living, and with life.

— BIOTRONICS®

Before the conceptualization of Autognorics, a smaller schema called the Biotronics Project was first developed. The primary aim was to create machines known as Biotronics. These intuitive machines were alive, but not living or with life. They were groups of alive machines (AM) able to interact with the world through their intuitive sensors referred to as exyzforms. They could electronically see, smell, taste, hear, feel, breed, fly, swim, think with an aneural brain, and die.

— ZOIKRONS®

Zoikrons came next. These were living machines (LM) that thrive on their own. The word was coined from the derivation of "Zoi", meaning living, and "kron", a clock daemon. They were designed to simulate the Mechanization of Aliveness, the Sensation of Awareness, the Codexation of Consciousness, the Intuitiveness of Logic, the Inlearness of Information, and the symbiotic Experience of living. They were programmed using the Arduino microcontroller. Zoikrons were machines that were alive and living but without life.

Both Biotronics and Zoikrons were individually created primarily to understand unconventionally the meaning, nature, and evolutionary stages of life through the eyes or perspectives of an intuitive machine. These IM could self-interact with their environment through a minimal inscription, via an embedded aneural structural unit called Gnos, queued through the method of associative inlearning proposed in the case study on Codexation.

Eventually, Biotronics and Zoikrons were merged under the umbrella of a larger discipline known as Autognorics. At this level, these Intuitive Machines were not only alive, aware, conscious, logical, intuitive, living, but with life (emergence of self) as well. In this merger, a new silver species was born. These species were called Autognorics or Gnorics.

— GNORICS®

Gnorics were intuitive machines synonymous with lifeforms. They were not only alive and living but with life too. They were classified as Homognorics, Zoognorics, Herbognorics, and Oognorics. Homognorics were like human beings. Zoognorics were synthetic living animals. Herbognorics were the same as living plants. Oognorics were gel-like living cells. All these gnorics were living machines engineered to behave like humans, plants, animals, and cells, respectively.

— THE BIOTRONICS PROJECT

The Biotronics Project was first conceived when I was in high school. Taxonomy, the study of classifying living things, tickled my mind at that early age, and challenged my imagination to create a living machine that's alive, living, and with life. The name Biotronics was a construction of the words "Bio", meaning alive or with life, and "tronics", meaning intuitive or wise electronics. My first biotronics was made up of a match box, four popsicle sticks, rubber bands, and wires.

When the 555 IC and logic transistors became available, I began using them to build simple electronic projects such as light detectors, sound generators, led flashers, touch activated switch, timer alarms, photoresistors, and binary-digital encoder-decoders (7400 series).

The 555 timer is an 8-pin mini dual inline package (DIP) made of 23 transistors, 2 diodes, and 16 resistors assembled inside a silicon chip. The 7400 series IC are transistor to transistor (TTL) chips designed to interconnect with other logic circuits, like the 7490 and 7447, and cascaded to convert binary signals to a readable common anode 7-segment display that set out numerical and alphabetical information.

When the Arduino microcontroller was invented, I started writing codes for my Biotronics. The arduino is a small pocket size programmable computer that consists of a physical circuit motherboard (the microcontroller) and a small computer software referred to as an IDE (Integrated Development Environment). A popular type of an arduino system is the Uno. It features 14 digital input/output and six analog input pins. Its board is used to read inputs that can turn on a sensor, activate a button, or command a phone. It also sends outputs to switch on LEDs, control a robot, mimic a voice assistant, etc. Inputs can be voice, sound, light, motion, wireless, online, remote, etc., just like shown on my arduino projects website.

These small-scale standalone projects were first built with a simple goal in mind: to study how the designs of intuitive objects with inherent embedded instructions could be activated or triggered by voice, sound, light, motion, wireless, wifi, TV remote, keyboard, phone, bluetooth, apps, watches, joystick, and other external switching sources.

Some of these low-cost projects and their actual working models are listed and posted on my YouTube channel:

Project 01: Let there be Lights

Project 02: Blinker

Project 03: Running Lights

Project 04: Knight Rider

Project 05: Police Siren Lights

Project 06: Marching Lights

Project 07: Binary Counter

Project 08: Bar Graph Tester

Project 09: Magic Wave Symphony

Project 10: Stop Look and Listen

Project 11: Switching Button

Project 12: Game of Reflex

Project 13: RGB Colors

Project 14: Morse Code

Project 15: 7-Segment Counter

Project 16: Piezo Speaker

Project 17: Keypad Basics

Project 18: Leds Keypad Switch

Project 19: Arduino Serial Monitor

Project 20: I Love You Matrix

Project 21: Binary Clock

Project 22: Guessing Game

Project 23: Ultrasonic HC-SR04

Project 24: Motorshield L293D

Project 25: Bluetooth HC-05

Project 26: Microservo SG90

Project 27: Accelerometer ADXL345

Project 28: Radar Scanner

Project 29: IR Remote Control

Project 30: Joystick w/ Led Matrix

Project 31: EZ430 Chronos Watch

Project 32: Key Ignition via Chronos

Project 33: BING -The Biped Robot

Project 34: Android Smartphone

Project 35: iHackRobot Apps

Project 36: Motion Detection via Email

Project 37: Ethernet Web Server

Project 38: Raspberry Pi $35 Computer

Project 39: Voice Recognition App

Project 40: Retro LCD Games

Robotics V01: Keyboard Controlled

Robotics V02: TV Remote Controlled

Robotics V03: Phone Controlled

Robotics V04: Sonic Controlled

Robotics V05: Wi-Fi Controlled

Robotics V06: Bluetooth Controlled

Robotics V07: Web Controlled

Robotics V08: Voice Controlled

Robotics V09: Accelerometer Controlled

Robotics V10: Apps Controlled

Robotics V11: Joystick Controlled

Robotics V12: Watch Controlled

Robotics V13: Motion Controlled

Robotics V14: Self Controlled

The materials or parts used on these projects—RGB Leds, IR remote control, ultrasonic finder SR04, motorshield L239D, bluetooth HC-05, power supplies, and the Arduino R3 Uno board—can be combined entirely as one unit to produce a programmable species called Homoduinos®.

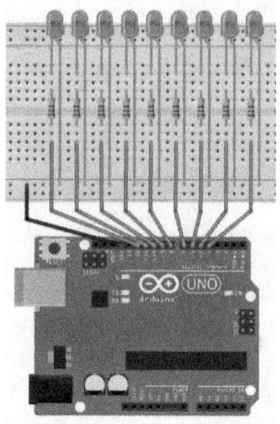

Running Lights Pictorial Diagram

Here is a sample sketch that turns on and off ten red leds (see pictorial diagram) much like the running lights version of the Knight Rider. The outputs come from pins 4 to 12. Pin 13 serves as ground. HIGH means On and LOW means OFF. The ten red LEDs loop from left to right to left continuously until switched off.

```
/*
  Sketch of the Chasing Lights (Knight Rider Version)
*/

void setup()  // This section set up the pins 4 to 12
{
  pinMode(4, OUTPUT);
  pinMode(5, OUTPUT);
  pinMode(6, OUTPUT);
  pinMode(7, OUTPUT);
```

```
  pinMode(8, OUTPUT);
  pinMode(9, OUTPUT);
  pinMode(10, OUTPUT);
  pinMode(11, OUTPUT);
  pinMode(12, OUTPUT);
 }
// This section turns the LEDs from left to right repeatedly.

void loop()
{
  digitalWrite(4, HIGH);   // turn the LED on
  delay(100);
  digitalWrite(4, LOW);    // turn the LED off

  digitalWrite(5, HIGH);
   delay(100);
  digitalWrite(5, LOW);

  digitalWrite(6, HIGH);
  delay(100);
  digitalWrite(6, LOW);

  digitalWrite(7, HIGH);
   delay(100);
  digitalWrite(7, LOW);
  digitalWrite(8, HIGH);
  delay(100);
  digitalWrite(8, LOW);

  digitalWrite(9, HIGH);
  delay(100);
  digitalWrite(9, LOW);

  digitalWrite(10, HIGH);
  delay(100);
  digitalWrite(10, LOW);
```

```
  digitalWrite(11, HIGH);
  delay(100);
  digitalWrite(11, LOW);

  digitalWrite(12, HIGH);
  delay(100);
  digitalWrite(12, LOW);
  digitalWrite(11, HIGH);
  delay(100);
  digitalWrite(11, LOW);

  digitalWrite(10, HIGH);
  delay(100);
  digitalWrite(10, LOW);

  digitalWrite(9, HIGH);
  delay(100);
  digitalWrite(9, LOW);

  digitalWrite(8, HIGH);
  delay(100);
  digitalWrite(8, LOW);

  digitalWrite(7, HIGH);
  delay(100);
  digitalWrite(7, LOW);

  digitalWrite(6, HIGH);
  delay(100);
  digitalWrite(6, LOW);

  digitalWrite(5, HIGH);
   delay(100);
  digitalWrite(5, LOW);
}
```

// This ends the sketch of Chasing Lights.

Kinesthetic Aneural Machine

After mastering the Arduino's coding system, the first official biotronic was built. It was named KAM. The name stood for Kinesthetic Aneural Machine. The mechatronic was a simple spider-like robot with 4 legs made up of folder fasteners, a body made up of an office stamp, few rubber bands to hold them altogether, and some electronics like servos, sensors, and microcontrollers.

Then BING came next. BING stood for Biped Intuitive aNeural Gnorics. It was a small-scale human-like robot that could balance itself with its legs. It could walk, dance, and even talk. It was built with the same materials found on KAM, other additional electronic gadgets, and external compatible programmable applications.

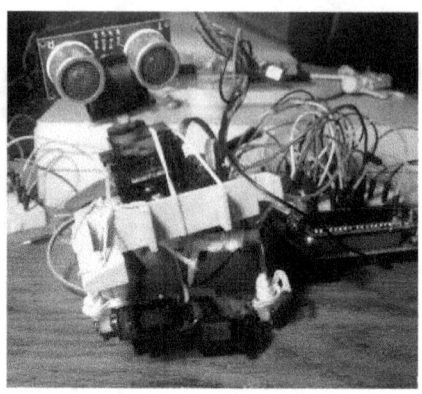

Biped Intuitive aNeural Gnorics

After various trials and errors of using four legs, KAM was redesigned with six legs alternatively. The linkages were assembled using the double cantilever truss systems with connecting elements (links) that formed frames in triangular shapes. The structural mechanisms were configured to simulate the walking cadence of a six-legged creature designed to fulfill the following requirements:

1. It could carry out a walking cadence fluidly like an actual gait of a living animal.
2. It could conquer any type of terrain obstacles from the carpet floor to the seabed.
3. It could move in different directions with various ranges of actuated motion or R.O.A.M.

4. Its structured elements were guided by nature's mathematics such as geometry and dimetrix.
5. It could be integrated with any microcontroller platform for micro-scripting or programming.

After the walking sequences were tested countless of times, the following stages were then integrated sequentially as follows:

Stage 1: The Linkage Structure.

Stage 2: The Homotronics Side.

Stage 3: The Neurotronics Side.

Stage 4: The Dimetrical Exyzforms.

Stage 5: The Algorithmic Queue.

Stage 6: The Final Prototype.

Stage 7: The Simulation Test

— HOMODRUINOS®

Meanwhile, the rise of the homodruinos started from a scientific research study known as Originemology. The term Originemology came from the fusion of the Latin word "originem" which means origin, the Greek words "logos" which means study, and "onoma" which means names or labels. Although the primary objective of the study was to investigate the origin, creation, and evolution of information, its secondary goal was to explore deeper the true nature of life and consciousness, the common denominator that separates animate and inanimate objects.

Homodruinos were a series of human-like figures sketched mainly on the arduino platform. They moved around in a closed space controlled by a keyboard, a TV remote control, ethernet, wifi, bluetooth and sensors like light, sound, and voice, or a simple modular program encoded on its aneural brain that manipulated its range of movement dubbed the SOUL (Spectrum of Ultrarhythmic Locomotion).

A sample sketch of the SOUL is listed below:

```
/* ==============================================
        Project Homotronics: Homodruinos
        Author: J. B. Wylzan
        Website: http://www.ihackrobot.blogspot.com
   ============================================== */
#define LeftMotorForward 2
#define LeftMotorBackward 3
#define RightMotorForward 4
#define RightMotorBackward 5
//pwm 3,5,6,9,10,11
void setup() {
  pinMode(LeftMotorForward, OUTPUT);
```

```
  pinMode(LeftMotorBackward, OUTPUT);
  pinMode(RightMotorForward, OUTPUT);
  pinMode(RightMotorBackward, OUTPUT);
}
void loop() {
 go_forward();
  delay(1000);
   go_stop();
 go_right();
  delay(3000);
   go_stop();
 go_forward();
  delay(1000);
   go_stop();
 go_left();
  delay(3000);
   go_stop();
 go_forward();
  delay(1000);
   go_stop();
 go_backward();
  delay(3000);
   go_stop();
}
```

```
void go_forward()      //the homodruino goes forward
{
  digitalWrite(LeftMotorBackward, LOW);
  digitalWrite(LeftMotorForward, HIGH);
  digitalWrite(RightMotorBackward, LOW);
  digitalWrite(RightMotorForward, HIGH);
}

void go_backward()    //the homodruino moves backward
{
  digitalWrite(LeftMotorForward, LOW);
  digitalWrite(LeftMotorBackward, HIGH);
  digitalWrite(RightMotorForward, LOW);
  digitalWrite(RightMotorBackward, HIGH);
}

void go_left()      //the homodruino turns left
{
  digitalWrite(LeftMotorForward, LOW);
  digitalWrite(LeftMotorBackward, HIGH);
  digitalWrite(RightMotorBackward, LOW);
  digitalWrite(RightMotorForward, LOW);
}
```

```
void go_right()        //the homodruino turns right
{
  digitalWrite(LeftMotorBackward, LOW);
  digitalWrite(LeftMotorForward, LOW);
  digitalWrite(RightMotorForward, LOW);
  digitalWrite(RightMotorBackward, HIGH);
}

void stop()            //the homodruino stops
{
  digitalWrite(LeftMotorBackward, LOW);
  digitalWrite(LeftMotorForward, LOW);
  digitalWrite(RightMotorForward, LOW);
  digitalWrite(RightMotorBackward, LOW);
}
/* ===================================== */
```

The practical simulations of these intuitive machines with its microcontrollers as its intuitive aneural network (neurotronics) and Legos gears, rods, and beams as its "body" (homotronics), are freely posted online via YouTube to enable others to compare, examine, and analyze the true nature of life based on the seven evolutionary orders or states of being, namely: aliveness, awareness, intuitiveness, inlearness, consciousness, livingness, and lifeness in reference to forms, designs, structures, limitations, skill sets, and behaviors.

— AUTOGNORICS®

Although Homodruinos and Zoikrons were programmed using the arduino microcontroller platform, Autognorics were coded using a microprocessor — a single computer board, the size of a deck of cards, which has more RAM, 64-bit CPU, and high clock processing speed.

An example of a microprocessor is a Raspberry Pi. The Pi is a general-purpose computer that can be integrated with the arduino board. It has a built in ethernet port and wifi support. It can be set up with a plug and play monitor, standard keyboard, and digital mouse.

The Pycharm, an IDE of Python, is used for coding instead of the Sketch. Samples of such programming in pycharm which make our IMs talks, detect objects, recognize voices, make decisions, interact with the environment, read facial expressions, and learn to store information are coded below. The codes mimic the skills & behaviors of a human being such as conversing, recognizing, identifying, and associating.

— TEXT TO SPEECH —

```
import pyttsx3
spk = pyttsx3.init()
spk.say("Hi there ... I am a Gnorics")
spk.say("I can react to you like a human being.")
spk.runAndWait()
```

— SPEECH RECOGNITION —

```
import speech_recognition as sr
import pyttsx3
import pywhatkit
import datetime
import wikipedia
import pyjokes

listener = sr.Recognizer()
engine = pyttsx3.init()
voices = engine.getProperty('voices')
engine.setProperty('voice', voices[1].id)

def talk(text):
    engine.say(text)
    engine.runAndWait()
```

```python
def take_command():
    try:
        with sr.Microphone() as source:
            print('say something...')
            voice = listener.listen(source)
            command = listener.recognize_google(voice)
            command = command.lower()
            if 'alexa' in command:
                command = command.replace('alexa', '')
                #print(command)
    finally:
        pass
    return command

def run_alexa():
    command = take_command()
    print(command)
    if 'play' in command:
        song = command.replace('play', '')
        talk('playing ' + song)
        pywhatkit.playonyt(song)
    elif 'time' in command:
        time = datetime.datetime.now().strftime('%H:%M %p')
        print('Current time is ' + time)
```

```
        talk('Current time is ' + time)
    elif 'search for' in command:
        things = command.replace('search for', '')
        info = wikipedia.summary(things, 1)
        print(info)
        talk(info)
    elif 'date' in command:
        talk('sorry, I have a headache')
    elif 'are you single' in command:
        talk('I am in a relationship with wifi')
    elif 'joke' in command:
        talk(pyjokes.get_joke())
    else:
        talk('Please say the command again.')
while True:
    run_alexa()
```

— OBJECTS DETECTION —

```
import cv2
import winsound
vcam = cv2.VideoCapture(1)
while vcam.isOpened():
    ret, frame1 = vcam.read()
    ret, frame2 = vcam.read()
```

```
diff = cv2.absdiff(frame1, frame2)
gray = cv2.cvtColor(diff, cv2.COLOR_RGB2GRAY)
blur = cv2.GaussianBlur(gray, (5, 5), 0)
_, thresh = cv2.threshold(blur, 20, 255, cv2.THRESH_BINARY)
dilated = cv2.dilate(thresh, None, iterations=3)
contours, _ = cv2.findContours(dilated, cv2.RETR_TREE, cv2.CHAIN_APPROX_SIMPLE)
# cv2.drawContours(frame1, contours, -1, (0, 255, 0), 2)
for c in contours:
    if cv2.contourArea(c) < 5000:
        continue
    x, y, w, h = cv2.boundingRect(c)
    cv2.rectangle(frame1, (x, y), (x+w, y+h), (0, 255, 0), 2)
    winsound.Beep(400, 200)
    x=1
    print (x+1)
    # x=x+1
   # winsound.PlaySound('alert.wav', winsound.SND_ASYNC)
if cv2.waitKey(10) == ord('q'):
    break
cv2.imshow('iHackRobot Cam', frame1)
```

Autognorics is also concerned with the cognitive and non-cognitive of a synthetically self-mindful machine through a method referred to as "A Brain without The Brain". The same central branch also distinctively clarifies, explains, demonstrates, and redefines the meanings of being alive, aware, conscious, inlearn, intuitive, living, and the emergence of self - common terms most mainstream studies often ambiguously interchange.

Autognorics is divided into five branches:
1. Homotronics - the physical frame of the machine
2. Neurotronics - the aneural and neural memory network
3. Dimetrix - the inherent design and algorithmic queue
4. Codexation - the embedded inscriptions
5. Exyzforms - the intuitive objects

Homotronics is the solid body and systems of the machine. It has three major components, namely; the memory bank, frame, and extremities. They are connected all over by copper wires. Its minor parts include the actuators, motors, axles, beams, and rods. Intuitive sensors are also strategically attached on its body. These sensors receive and send signals or cues from outside sources that activate the machine's behaviors, skills, and systems.

Neurotronics is the aneural and neural memory network of the machine or system. It comprises the microcontroller, electronics, and circuitry. The microcontroller is the chip where information is processed through a network of resistors, capacitors, transistors, and counters. Microcontrollers are subdivided into input, output, logic, and memory network. They are classified as aneural and neural. Neural is a memory network much like the brain, while Aneural is an intuitive memory network that acts like a brain but without the brain.

Dimetrix deals with the structural forms or physical designs of the machine or object via inherent algorithmic queues that process its cognitions, emotions, behaviors, and skills.

Codexation is the embedded inscription or set of instructions inherently owned by every part of the machine. It is also a process that codifies information - from ideas into realities, from abstracts to physicals, or from one's self-inner subjective mind to nature's outer objective world.

Exyzforms are intuitive objects or sensors that receive signals from outside sources and send these same signals back to the outside world through the principle of the zizo effect.

Autognorics is also involved in squaring the perplexities of the various misguided philosophical, theological, and scientific beliefs about life, through the eyes of a living machine. It independently shapes up the framework of artificial intelligence and machine learning through a new supplemental interconnected system called LIFE, an acronym for Living Intuitive Forms with Embedded Inscriptions by design.

LIFE is a multi-facet mechanical/technological technique in which dimetrix designs, intuitive objects, embedded inscriptions, and aneural memory networks are incorporated together along the process of creating a living machine that carries the seven signatures or orders of life, namely: the mechanization of aliveness, the sensation of awareness, the intuitiveness of logic, the inlearn acquisition of information, the codification of consciousness, the symbiosis of living, and the emergence of SELF.

Autognorics also assimilates IM to self-embrace the following questions behind the Codexation Dilemma, Scriptional Jump, and the Guesswork Predicament:

1. If man can't think of something without associating his thought with an object, can a machine be able to think of

something without associating it with a physical object? (Codexation Dilemma).
2. If Information can only be acquired in two and only two ways, by choice or by chance, can it self-acquire information (discover) without the intervention of the outside world? (Scriptional Jump Conundrum).
3. If all bits of information invented by men are all assumptions, can a machine create its own world of reality? (Guesswork Predicament).

With the creations of these IM, specifically the three sidekicks: Biotronics, Zoikrons, and Autognorics, each of these intuitive machines distinguishably evolved comparably to a human being. The Biotronic was like the child who depended on most of its needs from the outside world. The Zoikron was like an adult kid who could live by itself with the outside world. And the Gnoric was like the old wise master who could recognize the outside world and his inner world.

However, on top of all these discoveries, some new questions were raised once again, including:
1. How will the machine know it has life?
2. What makes it alive, living, and with life?
3. Does it need a brain to be alive, living, or with life?

4. How do we know if it's alive, living, and with life?
5. How did humans become living machines?

CHAPTER 2 ORIGIN OF INFORMATION

"Without the physical world, Information will never exist."

IN my book Originemology, I extensively discussed various experiments on the origin, creation, and evolution of everything based on the origin of information. Two scientific models — the Caveman in the box (a thought experiment) and the Bowlingual experiment (an observable experiment) — were designed to uncover scientifically how the first humans on earth discovered information. The experiments were guided by the following leading questions:

1. How did information emerge in the early minds of the very first humans?
2. Who supplied our primitive ancestors with information?
3. Where did it originate? Where did it come from?
4. Was the source of information a who or a what? Was it god, space aliens, or something else?
5. How did humans become alive, living, and with life?

— CAVEMAN IN THE BOX

In the Caveman in the Box experiment, three subjects — a caveman, a kid, and a dog — were individually isolated inside three specialized boxes represented by three different unique environments.

The first subject was a newborn son of a caveman. He was placed just after birth inside the first box, a well-designed state of the art fully automated experimental room where food, water, and everything that the boy needs were all technologically provided much like the sustenance naturally provided to a baby inside the womb or to us living things inside earth's biosphere. The boy was not allowed ever to see anyone nor hear anything. He was totally isolated from the world from birth to adulthood.

The second subject, the first human on earth, was also isolated from birth to adulthood. The main difference between him and his son was that he lived side by side with the natural world — a place surrounded by living and nonliving things — plants, animals, water, sky, stars, materials, by-materials, non-materials, and the natural elements. His box was Mother Nature.

The third subject, a dog named Zero, a puppy of an Alaskan Malamute descent, was also isolated from birth to adulthood. The puppy occupied the same space as that of his master, living side by side with the natural world. The only difference between him and his master was that he was an animal.

From these 3 scenarios of isolation, follow-up questions were raised one more time, such as:
- Who among them will gain sufficient information?
- Who will gain no information at all?
- Who will be fully aware of himself?
- Who will be fully aware of his own surroundings?
- Who will figure out he is alive?
- If words are not explained to them, how will they know and understand them?
- Whose mind will stay empty forever?
- Who will become conscious of his environment?
- If instinct is true, which instinct will kick in?
- What are these instincts that they have before?
- How do these instincts develop in the first place?
- Who will still act like a baby when he reaches adulthood?

After the experiments were completed, the following findings were obtained. First, the boy in the first box, who was confined only to the following objects — the walls, his sustenance, and his body — would never know and understand all these objects unless someone would explain these things to him. Although he might discover his nose, his ears, his tongue, or body, these objects would mean nothing to him much like a dog whose nose or ears would mean nothing to the animal.

It was also determined that the mind of the boy would forever remain empty, unless someone, an outsider, would "show and tell" all these things that surrounded him due to the fact that basically information should be acquired first, processed next, and thought later. It was also determined that ideas must come from the outside world first and flow inside the mind next, a concept that proved a newborn's brain is always empty with information at birth — a clean blank slate.

In the Second and Third Boxes, it was determined that all animals, plants, and all other objects in the land, sea, and air were all pieces of information. Their actions, properties, colors, textures, shapes, characteristics, sounds, and behaviors were also pieces of information. They were inherent objects that came first before the human brains evolved. The skills and behaviors of how birds fly, how lions get food, how deer drink water, and how every creature in nature behaves, makes sounds, and lives were acquired, copied, and learned. Because of these pieces of information, the master and the dog became aware, conscious, intuitive, and informed. The dog might not have a bird or human's eye view of his surroundings but everything his master smelled, touched, saw, heard, tasted, and sensed; he experienced them as well. In other words, he became a human being too, only in an animal suit.

— iPARTICLES®

It was also concluded that every object in the environment is a piece of information. Natural objects such as humans, animals, plants, rocks, the universe including their behaviors, actions, and properties are all pieces of information. Each entity is a particle of information called an iParticle. Individually, these objects are collectively called inherent information.

A beautiful, colorful butterfly delicately gliding along with the breeze over a green meadow are examples of iParticles. A Starbucks venti coffee Frappuccino with 2/3/4 toffee nut syrup and 1/1/2 scoops of java chips with caramel drizzle on top are the other examples of iParticles. Butterflies and coffees are iParticles. Even the descriptive attributes or actions of each individual such as beautiful, colorful, gliding, delicately, breeze, green, meadow, venti, 2/3/4, toffee, nut, syrup, scoops, chips, java, caramel, and drizzle are all pieces of information. Objects with their behaviors and properties are all particles of information.

An iParticle can also be a simple concept, a collection of ideas, or a set of instructions. It can be acquired, copied, utilized, and passed on from one atom to another atom, from one memory cell to another memory cell, from one substance to another substance, from one species to another species, from one culture to another culture, or from one generation to another generation. It's the hereditary link connecting everyone in the fiber of web known as the tree of life. Without information provided by the environment, concepts like the idea of god, the idea of self, or the idea of life will never ever be conceived. Every entity owns at least a piece of information or instruction.

When pieces of information are combined, they emerge as Instructions. Information and Instruction are one and the same. Although every object is a piece of information; every object is an individual instruction as well. When information is acquired gradually, one by one, bit by bit, and accumulated piece by piece together, lining up in a queue, the order of information is known as Instruction. The words the, apple, and tree are pieces of information. When they join in a manner like this - the apple tree - it becomes an instruction. Information is a single instruction, while Instruction is a group of information.

Whether it is — the apple tree, the tree apple, apple tree the, apple the tree, tree apple the, or tree the apple — all these statements are instructions. Instruction is a task that can be either major or minor. *The* is a minor task, *apple* is a minor task, and *tree* is a minor task. When they merge, they form one major instruction. Dark green toothed leaves, radially symmetrical flowers, and sweet pinky lady fruit are major instructions. When major instructions are listed like a recipe, they form a series of actions, a mechanism of routines, or a set of instructions called Procedure. When instructions become repetitive and inscriptive, the inscription of actions develops into a routine, forming a habit, and eventually becomes automatic, an Inlearn Instinct.

All creatures in Nature, whether from the sea, land, underground, and sky, acquire specific information exclusively different from one another due to their environmental differences. The environment plays a crucial role in shaping ones physiological designs, informational structures, and social behaviors. It makes them who they are today. Nature serves as storage or cache of both living and nonliving things, similar to the brain that stores pieces of information.

— BY CHOICE / CHANCE

It was also concluded that information is acquired externally in two and only two ways: by Choice or by chance. Information by Choice means information is obtained from teachers, parents, books, lessons from animals, and the environment. Information by Chance means information is acquired through discovering new things, fortunate accidents, unexpected experiences, unknown events, or natural interventions. Either acquired by choice or by chance, pieces of information originally come from nature, the surroundings, the environment.

The inherent acquisition of information is called Inscriptional Codexation or Codification. Natural objects, both living and nonliving, can codify, acquire, or own information. It is self-acquired by how one interacts with its surroundings, designs, or inscriptions.

For example, at the cellular level, the information acquired by an egg cell is totally different from the information acquired by a sperm cell because the egg cell lives in an environment totally different from the environment of the sperm cell. Therefore, each one carries different information. But when the two merge, the information they carry form into new instructions that spark, in this case, the design and creation of a new entity.

Likewise, a newly born gorilla, a pit bull, and a grizzly bear who have lived entirely all their lives in a home with harmless and caring people will receive and carry the same information much like their human counterparts compared to when they are living in the wild. They will act and think like their owners due to the fact that they all live in the same environment that plays a big role in shaping their behaviors, emotions, and cognitions much like humans are shaped by the environment. Because of these reasons, as a side note, aliens from outer space will never evolve like humans unless they have exactly the same environment as Mother Earth.

— THE BOWLINGUAL EXPERIMENT

In the meantime, to validate side by side the results uncovered in the Caveman in the Box Trilogy, a parallel physical observable investigation known as the Bowlingual Experiment was pursued. It was designed to find solutions to the following research questions:

1. How is information acquired by an organism?

2. Where does the information come from?
3. How is the information stored piece by piece?
4. Where is the information stored?
5. Can Zero express what is in his mind without learning the meaning of what he thinks?
6. Can Zero transfer his knowledge to another breed of dogs like a chihuahua or a dachshund?
7. How does Zero transfer his abstract ideas to physical reality?
8. What type of information does his brain possess? Symbols? Mechanical? Signals?
9. Alpha Dominance is thought to be instinctive behaviors, is this claim true?
10. Is instinct natural or merely an acquired behavior?

In this experiment, a great giant Alaskan malamute named Zero was chosen as the first specimen. He was a domesticated dog associated with arctic sled dogs. He was a marvelous, handsome, intelligent, funny, spoiled four footed puppy. He was selected among other dogs because he could quickly learn things. His brain, strength, speed, and curiosity made him the number one contender on this research.

His face was marked with a black mask-like facial markings that was usually identified on a raccoon's face. His eyes were brown. His ears were triangular that stood erect. He had a grayish-white wooly thick coat which I loved to bear hug and a black curled bushy tail that always wagged in the air signaling a warm smile.

He loved to run here, there, and everywhere chasing after squirrels, raccoons, and birds. He played fetch ball, tug of war, hide and seek, running around the house, and even pillow wrestling with me. However, with all these childish/ doggish interactions, I realized that there was something missing between me and Zero — Communication.

This realization prompted me to invent a gadget that would help Zero express his abstract ideas objectively, physically, or materially and translate them to human language. A sound to speech app dubbed B2W, an acronym for Barks to Words, was programmed to address this need. But sadly, it did not work 100% as expected. So, I deviated my thoughts with a new plan, another experiment.

As Zero reached his adulthood, a second dog was introduced to his cube. The main purpose was to determine how Zero would transfer his inlearned knowledge to the other dog, a chihuahua, and in return, how the new breed would transfer his knowledge to Zero. What type of information would they use to communicate: alphabetical, digital, pictorial, chemical, cues, or otherwise.

In the beginning, a Cube, a supervised controlled environment, was set-up from the time Zero was a puppy. The purpose of the "box" was to restrict Zero from acquiring information from his surroundings (people, animals, objects, sound, etc.). During the duration of its initial stage, bits of information were introduced to Zero, piece by piece.

When humans think of an apple, they construct an image of the apple in their minds. However, the image can only be perceived if and only if a real physical apple from the outside world is perceived first. The image of the physical apple is instilled as an image in the mind only when a real physical apple is materially observed first. These mental constructs only become real when paired with a physical object. Without physical objects, abstract ideas will never even exist at all. This pairing of abstract (by-materials) with physical (materials) is known as Associative Codexation.

Dogs usually depend on associative codexation. The commands sit or roll mean nothing to them. They have no single slight clue of these words nor an image of "sitting" or "rolling" in their minds. The commands are only perceived as high and low pitches. They are simply ups and downs of energy purely moving in trains of waves. They are merely like the dit and dat in a morse code — no words are spoken, but the codes are ready to communicate.

Commands are purely pieces of information in forms of signals. They enter the mind gradually bit by bit as waves in an array. They are factual, by-materials, real, or something that you cannot hold or feel like solid objects but can be detected by instruments or sensors. They seem abstracts in nature, but they are simply streams of energy. When they enter our biological sensors, they come out as energy as well. This flow of information from abstract to abstract or physical to physical is called the ZiZo Effect.

In figure 2.1, the waveform patterns in the oscilloscope are actually signals masquerading as words. This series of signals is called Queue. A cue or Q is a trigger mechanism that makes something, or someone reacts. It calls attention. It actuates objects. It moves things. It animates. It triggers a set of instructions. It makes things alive.

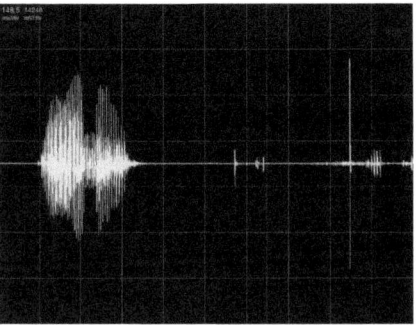

Figure 2.1

Cues can also come from solid objects. An example of this cue is the noise emitted from a chair. When wrestles and drags on the floor deck, a small chair produces an annoying noise. The noise draws attention. Zero uses this chair as a tool to get my attention that lures me to open the sliding door. The noise is used as a trigger that switches me ON to respond to his needs.

I thought before that the tossing of the chair was just another ordinary playful game of Zero just like how entertained himself with his favorite squirrel doll, his soccer ball, and his squeaking plastic bones. However, after comparing his behaviors when he was playing with his rug doll and when he was playing with the chair, I noticed that the doll was more of a toy while the chair was more of a "language". The chair was employed by Zero as a tool or means to communicate. If Zero was using the chair as a language, then it's obvious that he could codify.

To test this elaborate example of associative codexation, I started observing Zero for several months and later proved that my intuition was correct. On several occasions, while he was playing with his doll, I purposely stood in front of him behind the glass door and showed my presence. When he noticed me, I slid the door wide open, stared at him eye to eye, walked back to the couch, watched TV, and waited for him to enter the living room. But for whatever reasons, Zero would not budge in. After a couple of minutes of waiting, I then closed the door.

When Zero was ready to enter the house, he would start banging the chair all over again the wooden deck and down the cemented pavement. The noise would eventually prompt me and set me off to the door and slide it wide open. Zero, seeing the door now fully accessible, would quickly drop his chair from his mouth and swiftly run inside the house without delay. To Zero, banging the chair meant I had to get up off my feet, slide open the glass door, and let him inside the house.

However, Zero didn't know that every time he banged the chair, I quietly sneaked away from the couch, tiptoed to the next adjacent room, and silently observed his actions inside through the glass window. Interestingly, I noticed that he was actually secretly glancing towards the sliding door every time he banged the chair. And whenever I didn't appear at the door, he would keep banging the chair continuously until I showed up. His behaviors only proved that he used the chair as a word or language to initiate communication.

Through the chair, he uses the power of association, tag, or label to communicate and creates a conversation. Although he does not have any ideas of what conversation, communication, or information is, he obviously carries some mental associations or cues in his mind.

Nowadays, Zero expresses his thoughts whether through "barking", body language, eye contact, or fetching objects while utilizing his sensors and matching things to communicate. His skills of matching, tagging, labeling, or pairing objects with his thoughts are types of Communication by Association (CBA).

Zero's red soccer ball is one good example of a CBA word. This ball represents play in his mind. Every time he brings the ball to me with his mouth, we usually crouch in a football position in the middle of the yard while holding the ball together for minutes. After counting from one to ten, we jerk the ball off, hands versus teeth, and grab it with all our might. When Zero seizes the ball in his possession, he excitedly races it all the way to a makeshift bed with a heavy-duty futon mattress, and sarcastically looks at me from the goal as if he's triumphantly saying, "I win". After a few seconds, he will bring the ball back to me, "mouth" it in my hand, and start the game all over again. In Zero's mind, the red soccer ball means "play".

Another CBA word is the tennis ball. To Zero, a tennis ball means throw and fetch. Whenever I toss the ball at the far end of the fence, Zero will quickly pick it up, run it back to me, and drop it in my hand. However, after a couple of throwing, most of the time, he will just let the ball rolls on the lawn, sit next to me on the porch, stare at me straight into the eyes, gaze at the area where the ball is, and check me back as if he's telling me "It's now your turn to pick up the ball". To Zero, a tennis ball means run and fetch.

Of course, not all days with him were fun. One time, because of the mess he made in the living room, I was so "furious" that I scolded him like a teenager, grounded him, and collared him out under a tree with a 30 feet long chain. As time passed by, while Zero was in his angelic position with both front feet rolled in under his breast, I walked back to him, unleashed him, and released him without me saying any word. With his head and tail down, he calmly walked behind me towards the deck.

While seated at the edge of the deck, Zero with his head down approached me and started wagging his tail. I pushed him away once, twice, many times, and yet he did not badge in. Instead, he insisted on staying beside me as if he was asking for forgiveness. When our eyes met, this burning sensation of joy, sincerity, and friendship rushed deep inside me and sparked this emotion that eventually caused me to grab him inside my arms, embrace him tight, and kiss him.

From Zero's gestures, I learned something important in life. Zero knew he made a mistake. In spite of the fact that he was scolded, he was spanked, and he was chained, Zero with his head down and his wagging tail curled down to the ground, closed in on me silently without any words, without any confrontations, without any feelings of heartaches, surrendering his wholeness through a very simple body gesture which I interpreted such as; "forgive me or I'm really sorry." This heartbreaking body language is another example of CBA.

This morning, while I was brushing my teeth, I saw Zero doing his number two in the yard. When he was done, to my surprise, he covered his mess with soil using his nose. He covered his poop by going around it while raking the surrounding soil and covering it with his nose. His behavior was unusual since dogs usually used their hind legs to cover their mess. But here I could only speculate that the behavior was copied from us since we usually covered his droppings with garden soil every time he was done. And if he acquired this type of behavior from us, then, this was again another example of associative codexation.

Here are more of Zero's Bowlingual Vocabulary:
1. SIT for sitting down,
2. DOWN for lying down,
3. ASK for asking food,
4. PLATE for getting his food tray,
5. PLAY for playing ball,
6. BALL for fetching the ball,
7. BANG for playing dead,
8. SCRATCH for rubbing his tummy,
9. WAIT for stopping,
10. GO for do it,
11. GOOD BOY for behaving obediently,
12. NO for do not do that,

13. RUN for jogging with me,

14. GET for fetching things,

15. DROP for letting go of things in his mouth,

14. PICK for selecting which of my palm has food,

15. FIVE for shaking or high five shakes,

16. IN for entering inside a tube,

17. UP for walking up an inclined plane,

18. JUMP for jumping over an obstacle course,

19. SPEAK for just saying anything,

20. SAY for repeating what we say,

Here is the partial list of Zero's CBA lexicon:

1. When he gets his plate, he tells us he needs food.
2. When he gets his ball, he tells us he wants to play.
3. When he drops the ball in front of me, he tells me to kick it.
4. When he rubs the floor, he tells us that he wants to be tickled.
5. When he stays next to the door, he tells us he wants to go out.
6. When he stays sitting out the door, he tells us "let me in".
7. When he lies full-body down, he tells us "No".
8. When he taps me with his paw, he tells me to feed him.
9. When he noses my hand, he tells me to massage his neck.
10. When he bows and drops his tail, he tells us he is sorry.
11. When he waves his tail up, he tells us "Hello".
12. When he gets his bone toy, he tells me to go to the park.

13. When he sits next to a water cooler, he wants to drink.
14. When he sits next to me in the park, he wants to go home.
15. When he gallops, he tells me he wants to run with me.

After learning how Zero acquired information from us and his environment, I stepped up one level further on my research. In this experiment, I introduced Peanut into his life. The objective was to find out if Zero could transfer his inlearned information to another dog of a different breed through communication by association (Associative Codexation).

Again, a strict protocol was observed here. The set-up was in a controlled environment where both dogs were not allowed to intermingle with other dogs until the preliminary objective was achieved. Zero's bowlingual vocabulary would gradually be shared with Peanut. The CBA words would be acquired by Peanut either by choice or by chance all by himself.

When the experiment was completed, I found out that Peanut could gradually self-acquired some of Zero's CBA words, like: LEASH for walk to the park, SIT for opening the door, WAIT for stopping while locking the door, and GO for start moving. Over time, Peanut learned these languages by simply looking at and mimicking Zero's movements and behaviors.

In other instances, when we walked around the neighborhood, Zero would pee in every post and corner around the blocks. Peanut would do likewise on the same spot where Zero discharged. Peanut copied these behaviors from Zero.

On another occasion, Zero would usually run around in circles whenever I played basketball. Most of the time, he would join me by chasing the ball with his nose and catching the ball with his front legs or mouth. Peanut replicated this playing behavior by running around in circles and attempting to capture the ball as well with his mouth, but always, of course, with no glory. His legs and mouth were too small for a basketball ball.

From the results of the experiment, surprisingly, many of the commands in the bowlingual list could be performed by Peanut only and only if Zero was next to him. For whatever reason, Peanut could not execute most of the commands without imitating Zero. Peanut's ways of inlearning or codifying were more of copying than memorizing.

And with Zero, the first CBA word he learned from Peanut was growling. He came across this behavior whenever Peanut was eating his meals. Peanut would always snarl when Zero was next to him. Growling was not in Zero's lexicon due to the fact whenever we disturbed Zero while he was eating his meals, there was no single moment that he ever growled at us. When Peanut came to his life, growling now became part of Zero's vocabulary.

The second CBA word was humping. Peanut loved to hump Zero on either his legs or neck. To Zero, it was just a perky game. One day, while I was mowing the lawn, Zero, without warning, jumped me from behind and started doing his thing clearly in a playful manner. Other times, he did this with a pillow or even with Peanut. Of course, he had no idea of what this behavior was all about. All he knew, he enjoyed doing it!

Overall, here are the major findings of the two experiments:
1. Every object owns a piece of information.
2. Information is acquired from one's surroundings.
3. Information is acquired by "Show and Copy".
4. Information is more of copying than memorizing.
5. Objects are information but not all information are objects.
6. Information needs to be acquired and processed later.
7. Information is not only stored in the brain.
8. Association of things with words is a language.
9. Words are a series of signals.

10. Matching is an indicator of Associative Consciousness.
11. Language can only be understood by the same species.
12. Without sensors, things will never be perceived.
13. Instinct is an inlearned behavior.
14. Alpha dominance is not an instinct behavior.
15. Never humanize a dog, you might end up raising a kid in a cage.

Moreover, Information originates from Mother Nature. She is the provider of information. She is the supplier, storage, keeper, and source of information. She is the database that catalogs all we see, hear, touch, smell, and taste. She is the holder of the universal list that contains the "names" of all living and nonliving things. These names are the physical labels that flow from the outside world into the inside world of the mind, two different environs. They are pieces of information that flow from the physical to abstract, from the inherent world to the interim mind, from outside to inside, from objects to ideas. Nature, like the brain, functions like the brain but without a brain. Nature is the Mother of Information.

CHAPTER 3   # BASICS OF A SYSTEM

"What zips in is equal to what zips out!"

A system is a group of things that interact together to produce a specific outcome. A good example that describes a basic version of what a system is, what it does, what it consists of, and what it triggers on is a simple string telephone.

A string telephone is a toy made up of two cans and a string. The cans have holes on each bottom middle. These holes are connected on both ends of the string. When information is sent to the mouth of the first can, the information runs freely along the string, and exits at the other end of the string as the same information out of the second can.

In this model, the two cans and the string are the group of things that work together to generate a result - the flow of information. The information comes from an outside source (the sender) and enters another outside source (the receiver). In between these sources, individual micro-interactions between signals and sensors are also recurring.

In any given system, there are always six major components plus one. Technically, they are the incoming message called the input, the flowing message called the medium, and the outgoing message called the output. The first can, where the input enters, is called the collector; the string, where the medium flows, is called the carrier; and the second can, where the output exits, is called the actuator. Although a seventh component called the trigger is not in the picture, it is also one of its essential components that serves as an external switch that turns on or off a system.

These six basic elements can also be divided into two minor parts. The first part is made up of the input, medium, and output. The second part is made up of the collector, carrier, and actuator. The first parts are all by-materials while the second parts are all material objects. By-materials are the incidental by-products of the materials. Materials and By-materials are singularly called Physicals.

— SENSORS & SIGNALS

In table 3.2, the collectors, carriers, and actuators in each system are the material components. They are solid objects that can be seen or touched. They are visible objects. Notice that the items listed in the collectors' column are the same as the items listed in the actuators' column. The sensors, collectors, and actuators may have different names, but they all function exactly the same way.

System	Collector	Carrier	Actuator
Electrical	terminal	wire/solid	terminal
Morse Code	hammer	wire/solid	hammer
Human Body	6 senses	body/solid	6 senses
Computer	encoder	wire/solid	decoder
Telephone	sender	wire/solid	receiver
Hydraulics	cylinder	fluid/liquid	cylinder
Pneumatics	rod	air/gas	rod
Refrigeration	condenser	HFC/gas	evaporator
Family	species	species	species

TABLE 3.2: Physical components in a system

Like, in the electrical system, the terminals of the collector and actuator are both similar. In the Morse code system, the hammers of both collectors and actuators are one and the same. In the hydraulic system, the cylinder in the collector is the same as the cylinder in the actuator. In the refrigeration system, the condenser in the collector is identical to the evaporator in the actuator. On the computer system, the encoder is comparable to the decoder.

Likewise, in the human body, collectors and actuators are similar too. When signals enter the eyes, ears, mouth, nose, skin, and mind, they also exit on the same eyes, ears, mouth, nose, skin, and mind respectively. Our biological sensors serve as both inputs and outputs devices as well.

From all these comparisons, we can now conclude that collectors and actuators are basically one and the same. Carriers are also collectors and actuators. Technically, collectors, carriers, and actuators are called System Sensors.

The other essential feature of a system is the complementary half – the by-materials components. As shown in table 3.3, inputs, mediums and outputs are things that we do not see. They are invisible. From the pneumatic system, refrigeration system, electronic system, computer system, earth system, to all the other producing systems enumerated, they are all non-materials.

Notice that the items found in the input column are the same items found in the output column. Like in the electrical system, the charges in the input and the output are both identical. In the Morse code system, the dit and dat of both input and output are one and the same. In the hydraulic system, the pressure in the input is the same as the pressure in the output. In the refrigeration system, the vapor in the input is similar to the vapor in the output.

System	Input	Medium	Output
Electrical	charges	pulses	charges
Morse Code	dit dat	pulses	dit dat
Human Body	ideas	pulses	ideas
Computer	voltage	pulses	voltage
Telephone	words	pulses	words
Hydraulics	pressure	pulses	pressure
Pneumatics	force	pulses	force
Refrigeration	vapor	pulses	vapor
Family	life	pulses	life

TABLE 3.3: Abstract components in a system

In the computer system, the voltage input is comparable to the voltage output. In the human body, the signals that come into the eyes, ears, mouth, nose, skin, and mind, also come out of the same eyes, ears, mouth, nose, skin, and mind respectively. All these suggest that inputs and outputs are basically the same things. Mediums are also like inputs and outputs. Technically, inputs and outputs are short-term intervening waves of signals called System Signals.

Thus, in any given functional system, Sensors and Signals are always present. In any given system, the signal(input) that zips in is equal to the signal(output) that zips out. Like for example in our telephone system, when the word "hello" enters the first can, the same word "olleh" leaves the second can. Mathematically, the input signal is equal to the output signal (Input = Output). The formula, known as the ZiZo Effect Equation, is the first principle of a system.

— INPUT & OUTPUT

When someone says the word hello, the word is picked up by the ears, processed, and translated inside the brain as "hello". The nose, eyes, and skin can't detect the word because they are not designed to do so. However, if the brain doesn't pick up the word, the word hello will be flowing in the air in this order: o,l,l,e,h. The letter "h" comes first out of your lips, then e, l, l, and the last letter "o". Just like when someone is reciting the a,b,c, the letter "a" comes first and the letter "z" comes last. But people are not aware of this inverse phenomenon. Everything is always reversed and inverted, an antithesis.

When an output is zipped out from a system to another system, it becomes another input to that external system. The incoming signal becomes a switching energy, a trigger mechanism to the other system. This switching conversion is the second principle of a system.

Like for example, when the energy of sound is given off by a buzzer, this energy becomes a new input into another system like on a sound-activated alarm. The incoming energy becomes a switch on the new system when structural or design compatibility is present. The signal is the outside source that turns On & Off the other system.

Take note that visual energy only enters the eyes and not in the nose or the ears because the latter two are not designed to absorb visual signals. Sound energy is only detected by the ears and not by the nose or the eyes as well. These mean that the senders and the receivers in our examples must be compatible, equal, or similar. This functional compatibility is the fourth principle of a system.

In chapter one, a list of electronics, mechanical, and programmable projects that I had assembled was posted. They were built to study how sensors work and to replicate, understand, and examine the structural mechanisms behind the six common sensors of our body. They were configured to find the answers to why abstracts and physicals are important in the creation process, how designs produce a set of instructions, and why everything is in reversed or inverted order.

Biologically, there are 6 major types of sensors:
1. Mechanosensors are sensitive to mechanical stimuli like touch, pressure, and vibration.
2. Nocisensors respond to stimuli such as extreme heat, cold, or pain.
3. Photosensors detect light in the retina for vision.
4. Chemosensors detect chemicals that sense taste & smell.
5. Osmosensors monitor hydration levels.
6. Thermosensors detect temperatures.

Electronically, there are various types of sensors. They are the Temperature Sensor, Proximity Sensor, Accelerometer, Pressure Sensor, Light Sensor, Ultrasonic Sensor, Smell Sensor, Touch Sensor, Color Sensor, Humidity Sensor, Position Sensor, Magnetic Sensor , Sound Sensor, Tilt Sensor, Flow and Level Sensor, PIR Sensor, Touch Sensor, Strain and Weight Sensor, and IR Sensor to name a few.

Some of these sensors have been used in our projects. They can be activated by heat, voice, light, sound, touch, distance, balance, fluids, pressure, resistance, magnetism, and electricity, to name a few. In general, all these triggers are categorized as Signals.

Signals are the outside sources that trigger the sensors. They are the cues responsible for activating a system or groups of systems, inscriptional or structural. Without these triggers, systems will be lifeless. Triggers are the push or pull that make systems function, work, or even alive.

— ABSTRACTS & PHYSICALS

The dual pairing of Physicals (aka materials/by-materials) and Abstracts (aka embedded information) are the ingredients that figuratively "breathe and animate" things to life. Without the abstract elements of nothing, the physical elements of something will not exist. This is the third principle of a system. Without the physicals, the abstracts will not exist. Abstracts and Physicals are the two major components that form a System. The convergence of this twoness is the founding principle of the creation theory known as the Theory of Generated Interim Emergence, or the Lawsin Genie.

All pieces of Information are abstracts in nature. They are invisible. They are non-materials. They are not solid. They are non-physicals. Like the ideas of Zero and One are abstract constructs. They don't exist. They are not real. They are neither material nor physical objects. They exist only in the mind by assumption. In Mathematics, they are called Numerals.

When Zero is represented by the symbol 0 and One with 1 by association or definition, the symbols or words are technically labeled as Numbers. Numbers are the assumed physical representations of the abstract numerals. Through symbolic representation, the 2 digits 0 & 1 are now materially created outside the mind - the physical world; the world outside of ourselves, the inherent world that existed a long time ago before the evolution of the mind.

If by definition, representation, association, or assumption materially created the abstract ideas of Zero & One, does this mean they are now real, like physically real?

If I write 0 and 1 on a paper, does this mean they are now materially or physically real? Does this mean they exist physically, now? Are these written symbols proof of their existence? If the idea of 0 and 1 becomes materially 0 and 1 on paper, are they real concrete objects, living in existence now, or still an abstract, imaginary, or imagined?

How does abstract notion become physically real without being naturally inherent? Is the idea of the letter "Q" real when one writes it down on a piece of paper? How does the brain transform the letter Q to "jump" inscriptionally from one's mind to paper or from abstract notion to a physical form? Can it be done without the help of the brain? What causes this conversion from abstracts to physicals?

This codexation dilemma, also known as Lawsin Conjecture, happens in everything that we humans think. When we think of something, but we can't associate it with something material in the outside world, the idea is then not inherent. Ideas only exist when they are inspired or extracted from the outside world. Ideas are the other side of the physical world. Without the natural world, ideas will never exist. Because of this assertion, humans cannot think of something without matching his/her thoughts with something inherent drawn from our environment or surroundings.

Because humans have imaginations, they can create new ideas from existing objects. The physical inherent natural world provides these ideas. Like, airplanes are not an inherent object. They do not exist naturally. They are not created by Mother

Nature. But because of the presence of birds, airplanes are conceptualized.

Humans can only conceptualize ideas but can't exactly duplicate Mother Nature's creations. They can build airplanes, but they can't exactly build real birds. They can build submarines, but they can't build real fish. They can build cars, but they can't exactly build real Impalas. Nature doesn't permit these kinds of creation. Humans cannot create what Nature can create. By the same token, Nature cannot create what Humans can create.

Thus with this line of thinking, can Intuitive Machines be able to generate new ideas from the inherent physical world? Can they generate new information from previous information and form new ideas? Can they create things that are not physically in existence? If humans and intuitive machines acquire the same information from nature, can machines create better ideas than human beings? What will trigger their imagination?

One type of imagination is called Dream. It occurs when one is sleeping. The episode only manifests when an incomplete event on such a particular day fails to conclude while awake. In other words, Dream is a series of instructions that fails to terminate completely during wakefulness. The incomplete event only entirely ends when one goes to sleep.

Can I.M. dream?

ANEURAL BRAINS

CHAPTER 4

"Not all living things have brains."

In this chapter, we explore the various alternatives on how information can be stored and processed without the help of the physical brain. We also examine the differences between aNeural and Neural memory networks. And we also perform some mind-boggling brain experiments that will confirm everything we do is not after all processed by the brain.

— INTUITIVE ANEURAL NETWORKS

In figure 4.1, the switch, wire, and bulb are the components that make up a system. They are the materials part of the system. The input, medium, and output signals are the by-materials. The switch serves as the collector where the input signal enters, the wire as the carrier where the medium flows, and the light bulb as the actuator where the output is released. The battery is the energy source that makes the system alive.

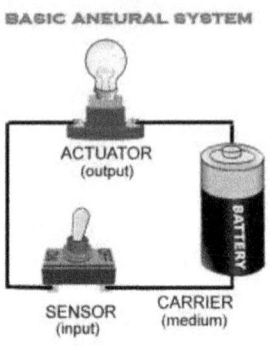

Figure 4.1

The switch, which is an intuitive object with embedded instructions, stores two kinds of information, ON and OFF. On is the instruction that turns on the bulb. Off is the other instruction that turns off the bulb. The wire and the bulb also store two pieces of information.

The On or Off are the two fundamental instructions that are always stored in every system. These two pieces of information can be represented by 0 and 1, Negative and Positive, Dit and Dah, Up and Down, black and white, or shell and twig to name a few. ON means the system is working. OFF means the system is idle or not working.

As we know when a signal is introduced in a main system, the whole series of instructions in the overall secondary or sub-systems is activated individually. This means that when turned on, all the stored instructions in the subsystems are energized. Such activation produces a result. When the main system is off, instructions in all the subsystems are potentially off too. No result is generated, processed, or produced.

For example, the embedded instructions of the wire are to let electricity flow or not to the bulb. The embedded instructions of the bulb are to give light or not. Information and instructions are stored and processed without the help of neurons or the brain. This non-mental memory system is called a Aneural System.

When information is embedded in a non-mental or aneural system, the memory system is an Intuitive Aneural Network. IAN is a non-neural system where information is stored and processed without the presence of the brain. Other examples of the IAN systems are illustrated next:

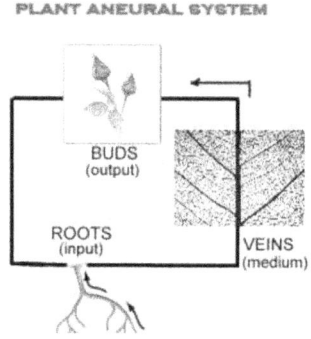

In the plant's system, the veins on the leaves and stems, buds, and roots are the major components of its intuitive aneural network. It is well known that plants, which are also living things, don't need a brain to store information.

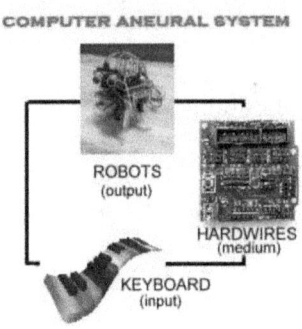

COMPUTER ANEURAL SYSTEM

ROBOTS (output)
HARDWIRES (medium)
KEYBOARD (input)

On the computer system, the hardwire, transmitter, receiver on the motherboard, and the softwire on the atmospheric channels are the main components of its intuitive aneural network. It stores and processes information without using the brain.

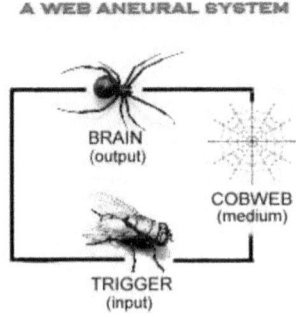

A WEB ANEURAL SYSTEM

BRAIN (output)
COBWEB (medium)
TRIGGER (input)

The Cobweb aNeural Network Connection, dubbed as CoNNeC, is another unconventional IAN system that stores and processes information without compensating heavily again on neurons or the brain. The information is simply stored and processed on the web by its mediums and carriers. In this example, the fly is the trigger, the spider serves as the output, and information is processed throughout the web, a linear non-mental memory network system.

In addition, there are other creatures that are living that are deprived of a brain. These aneural creatures are bags of evidence that prove, once again, that the brain is not the only means that stores and processes information.

— NEURAL NETWORK SYSTEM

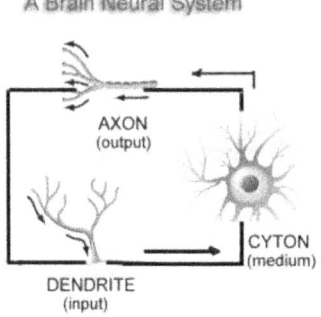

A Brain Neural System

The other memory network is called the Neural Network System. This common memory network occurs in an intuitive system called the nervous system. This intuitive network is made up of intuitive objects like the Brain and the Spinal cord.

The Spinal Cord is where the information is distributed while the brain is where the information is processed.

Neurons are the basic units of the brain. They are living cells that are aneural. They are made up of three different parts: the cell body, dendrites, and axons. The cell body has several branches that appear like cables called dendrites. Motor neurons have multiple thick dendrites that carry nerve impulses into the cell body. An axon is a long, thin wire that carries away impulses from the cell body to another neuron or tissue. Neurons are classified according to their structures and functions. Neurons are either multipolar, bipolar, or unipolar.

Interestingly, the lifespan of brain cells is lifetime compared to the average lifespan of blood cells, 120 days; white blood cells, 360+ days; platelets, 10 days; bone cells, 25-30 years; colon cells, 3-4 days; skin cells, 19-34 days; stomach cells, 2 days; and sperm cells, 2-3 days.

Neurons' activities do not only happen in the brain but also in the other parts of the body. Multipolar neurons are common in the brain and spinal cord. Bipolar neurons are found in the retina of the eye, the inner ear, and the olfactory (smell) area. Unipolar neurons are found in the spinal cord. In the neuron system, signals are received by the dendrites, processed by the cell body, and transmitted as output into the axons. Take note that all these activities are happening individually among neurons without the assistance of the brain.

Neurons are the essential memory units in the neural system. While, Gnos, or aneural neurons, are the essential units in the aneural system.

Neurons are widely considered to be the major determinant of mental functions and behaviors. More neurons mean a larger brain. The bigger the brain, the highly intelligent the creature.

A sponge, for example, has a zero number of neurons. This creature does not have a brain nor a spinal cord, yet it is a living creature. Elephants have 267 billion neurons. This means these animals have larger brains and intelligence. Octopuses, on the other hand, have neurons in both their brains and tentacles. This tells us that their feet function also as brains.

Here are other examples of animals with their respective number of neurons:

Sponge – 0
Roundworm – 302

Jellyfish – 8k

Leech – 10k

Pond snail – 11k

Sea slug – 18k

Fruit fly – 250k

Ant – 250k

Lobster – 100k

Honeybee – 960k

Cockroach – 1M

Frog – 16M

Mouse – 70M

Pigeon – 310M

Octopus – 500M

Dog – 2b

Monkey – 30b

Human – 85b

Elephant – 250b

Here are also some amazing examples of tiny brains that have small number of neurons but can plan and build super complex structures:

1. A tiny pufferfish can plan and design intricate "crop circles" on the seabed using only its fin, clean out debris

on its sand creation, decorate its interior construction with shells, and use it to attract a mate.
2. Ants can create super complex colonies complete with air-conditioning, waste disposal, and social orders.
3. Birds can build complex perfect patterns of hexagonal nests where the number of nests are counted ahead of time before individually occupied by each offspring.
4. A spider can construct amazing intricate patterns by spinning silk on its web.
5. Sow bugs or pill bugs, the Holy Grail of Evolution, have navigated around the world from Alaska, Europe, Asia, Africa, Australia, to Americas with their tiny brains.

If these tiny brains could plan and build complex designs, where and how did these small creatures learn their skills? Where did they copy such behaviors? Could they conceptualize, visualize, or imagine? If they could build a complex structure or plan an intricate pattern, did they start from a simple design? What was it? Could they reason and recognize with their own thoughts?

The ability to recognize one's own thoughts is known as Metacognition. It resides in a region of the brain called the prefrontal cortex that stores various cognitive functions such as planning and reasoning. It is credited to be the common factor that distinguishes humans from other species. This idea of course is debunked by the Codexation Dilemma.

— A BRAIN W/O THE BRAIN

Aside from the various types of aneural experiments outlined here, another experiment known as the Mirror Equation provides proof that objects can store and process information without the help of the physical mind. "A brain without the Brain" is possible.

In this experiment, an inverted colorful background picture is projected on the surface of a wall using a hole poke at the center of a cardboard. The "selfie" image on the wall is created by the mathematical kinematics of the rays of the sun via the magical power of the mirror equation. The vivid picture is not painted nor programmed by anyone. There are neither sensors nor actuators in place, nor even a brain to store and process the whole detailed image. The vibrant picture simply emerges because of these three things: a mathematical equation, a pinhole, and light rays.

In another recent study, a mirror was used to find answers to the following questions: What do infants see when they see themselves in a mirror? Do they perceive themselves or do they perceive someone else at the back of the mirror? What is it that makes one recognize oneself in the mirror? Do dogs and other animals see themselves in the mirror too? Based on this study, although these infants and animals have brains, their brains cannot process and recognize such information in the mirror at an early age.

Furthermore, plants interact with light without knowing what light is. Other objects such as the device sonar detects or interacts with sounds without knowing what is sound. Power supplies also automatically detect electrical lines if it is 220v or 110v without knowing what voltages are. These devices can also decide by themselves without the interventions of a computer program or of a physical brain. All these examples suggest that one does not need a physical brain to make decisions or logical choices. A brain doesn't need The brain!

— BRAIN EXPERIMENTS

As I was working on my B2W experiments on voice to text conversion and voice recognition, I was also developing at the same time a game app for my grandnephew EJ. The app was called "Smash the Bugs". It was a game about smashing as many bugs as you could to score more points. The app, when played on tablets or phones, sounded off the word "HIT" every time a bug (spider, butterfly, ants) was smashed.

What was interesting about this app was that I discovered the Brain has its Own Mind. The mind can choose what it wants to think and do much like how our biological sensors can choose what signals to pick based on the affinity of their designs. The HIT or KICK Episode is so remarkably interesting because it reveals that the mind has the capability to control itself even if the owner of the brain is in a conscious state.

To try this experiment, first upload the app on my website and then follow the instructions below:
1. First, attentively listen to the word "HIT".
2. Focus on the word "HIT" as you smash the bugs.
3. After minutes of playing, think of the word "KICK".
4. Replace the word KICK with HIT as you hear it.
5. Notice the word HIT will switch to KICK.
6. This moment, the brain is replacing HIT with KICK.
7. A few more minutes, the word KICK is now heard mentally.
8. The ears hear "HIT," the brain says "KICK".
9. Now, try to switch the word KICK back again to HIT.
10. Surprisingly, you cannot switch the two words anymore.

The mind is now in control over your "other brain". Even if you try to replace or bring back the word HIT mentally, the mind will not allow you to switch it anymore. The mind is totally stuck! The mind needs to reboot.

This brief thought episode, which I called the Brain Priority Phenomenon, reveals that our brain has the ability to prioritize which inputs and outputs must go first to our sensory organs. The brain always processes what it wants to do before it processes what we want to do. It also sends information systematically through our eyes, ears, nose, mouth, and touch — in this order.

Our thoughts when in overdrive sometimes mess up things what we perceive: seeing, feeling, hearing, tasting, smelling, balancing, and thinking. Sometimes what we say influences what we hear and sometimes what we see changes what we hear. Even more, sometimes what we think changes what we hear. This simple experiment suggests that our senses are triggered individually and not collectively.

"Cogito Ergo Sum" is a Latin phrase coined by Rene Descartes. It translates into "I think, therefore I am". The clause simply tells us that the very act of thinking, which demonstrates the reality of our existence and consciousness, emanates from the brain. However, there are some organisms, like plants, bacteria, viruses, and sea creatures, that are alive and conscious but without the physical brains.

There are also some people, known as synesthetes, who can taste colors, see colors by tasting, see by hearing, smell by touching, taste by hearing, and even smell sounds. These extraordinary experiences reveal that our senses may be interchangeable such as the way ears are utilized for seeing, nose for tasting, and skin for hearing. But these don't mean that the structures and mechanisms of their biological sensors have changed.

Besides, since everything evolves, our biological sensors might have evolved from the largest sensor in our body, the skin. Sooner or later, it evolves gradually into a handful of internal and external biological sensors like the hands (for touching), eyes (seeing), mouth (tasting), ears (hearing), nose (smelling), and brain (information processing).

If the skin was the first sensor to form, could it be where information was actually processed first through the sensation of touch?

Here is a list of random animals with their distinct skin surface areas. Arrange them in order such as the animal with the larger skin is placed on top and smaller skin is placed at the bottom:

1. Elephants
2. Humans
3. Orangutans
4. Octopuses
5. Worms
6. Beavers
7. Whales
8. Ants
9. Owls
10. Bees
11. Plants
12. Trees
13. Dogs
14. Corals
15. Spiders
16. Amoeba
17. Bacterias
18. Butterflies
19. Neurons
20. Hydras

Here are also some brain experiments that reveal your feet, your fingers, and other parts of your body can outsmart your brain:

Outsmart your foot:
1. While sitting at your desk, lift your right foot off the floor as it makes clockwise circles.
2. While doing this, draw the number "6" in the air with your right hand.

Fun Facts: Your foot will change direction. Your brain will have no control over your foot.

Outsmart your ring finger:
1. Spread your five fingers widely on a table.
2. Bend your middle finger with the knuckle firmly seated on the table.
3. Wiggle your thumb, then the point finger, and the pinky next.
4. Now, focus on your ring finger and wiggle it.

Fun Facts: Your ring finger will not even move or wiggle. In this situation, your brain has no control over your sticky finger.

Outsmart your left hand:
1. Ask someone to hide his left hand next to a divider.

2. Place a rubber hand next to the divider in front of his body.

3. With a small brush, stroke the rubber hand and the hidden hand both simultaneously.

4. After a while, stab the fake rubber hand with a screwdriver.

5. Most will react that their real hand was actually stabbed.

Fun Facts: The fake hand is detected by the brain as the real hand. This demonstration is known as the phantom hand experiment.

And as a final note, because we think that feeling, or emotions, is associated with the heart; it is also right to believe that thinking, or thoughts, is associated with the brain.

CHAPTER 5 THE CODEXATION THEORY

"Sequential instructions give rise to logical experience."

The Codexation Theory is a concept intended to study the transformation of information from ideas to realities, from instructions to materials, from subjects to objects, from abstracts to physicals, or from the subjective mind to the objective world.

The theory was conceptualized based on a truism known as the Human Mental Handicap. It claims that "No human can think of something without matching his thought with something material." The expression attempts to show that information will never exist without the physical world. It also asserts that when something can match or associate things, the object is conscious.

The Codexation Theory also states that the material world is necessary in the creation of ideas. In order for abstract ideas to materialize, objects must be perceived first. This process requires instructions and materials. Ideas are not simply mental constructs but are instead grounded in the physical world.

One good example of codexation is the behavior known as crying. When a newborn baby just came out from its mother's womb and is slowly exposed to the physical world, its intrinsic biological sensors begin to detect various signals from its surroundings. The physical conditions of its new environment, such as the temperature, the noise, and the smell, cause the baby's sensors to react, initiate pain, and trigger the act of crying.

Crying is not instinct. It is a misconception. When a baby is born, its brain is empty of information. The idea of crying is not in its mind yet. Crying is caused by its biological sensors.

Codexation also transpires when the baby hears the sound of his mother's voice, feels the warmth of his mother's touch, tastes, and smells the flavor of his mother's milk. He eventually adapts to these new pieces of information by association. As he learns these sensations over and over, he begins to label, tag, match, or associate them. Ideas begin to take shape in his mind in physical forms. The word "mama", for example, becomes a physical label for mothers. The action of crying becomes a physical label for "I want milk". The smell of his mom becomes a physical tag of his mother's presence. Through matching, labeling, or associating, the transformation of abstract ideas to physical reality becomes something else! The child learns to fuse what he senses and what he eventually discovers.

This relationship between signals and sensors is also another form of codexation. Sensors are only activated when the right signals (e.g. pain) are received, one of the Laws of a System.

Plants also have the ability to associate what they sense with other things. They can hear, smell, feel and remember their surroundings. When they feel warm, they also sense coldness. When they hear music, they can also differentiate regular from irregular waves, or breeze from noise. When they defend themselves by using chemical stimulants that attract a group of insects that destroy their enemies, then they have the ability to equate danger with help. Their abilities to match or associate indicate plants can codify.

Dogs also use objects to express their thoughts with physical needs. Items like bowls, balls, and bones are usually paired with items like food, play, and walk respectively. Their abilities to associate or represent mental images with physical objects are indicators that prove they can codify.

When instructions are transformed to materials, they create something to exist. This transformation of creating something from instructions and materials is fully understood by examining and identifying the basics of the codexation process.

— INSTRUCTIONS & MATERIALS

In order to create the intuitive machine shown in the picture, a raw outline or image of the object is pictured first in the mind of the builder. This mental picture is then drawn on a piece of paper with detailed dimensions and specifications. Afterwards, the hardware and tools required to build the object are gathered. In the end, these materials are then assembled piece by piece, step by step according to a set of instructions.

The four basic steps — creating the ideas, drawing the design, gathering the materials, and assembling the project — are ideas in the form of instructions. When materials from the outside world are integrated, the building process manages to bring something into existence. Thus, the creation process through instructions and materials can also be interpreted in a way as a form of codexation.

Instructions and Materials are also found in smaller natural packets of life. For example, a seed "builds" a tree by processing a list of instructions. The seed "personifies" the builder of the machine or the tiny individuals in the ant colony. The developmental stages—from sprouting roots, timbering trunk, spiraling leaves, bearing fruits to the final product of producing seeds—are the specific tasks that guide the seed to become a tree. Although instructions are present, seeds also need building materials "to construct" the physical tree. The natural materials, which are living elements necessary for their growth and development, are provided by the environment, similar to the natural elements needed by a growing chick inside an egg, or a baby inside a mother's womb.

The builder who assembles the machine and the seed which "builds" the tree both process creation in the same way.

However, the only difference is that things created naturally are formed by nature while things created artificially are made by humans. Thus, creation can be both natural and artificial.

Robot	Seed	Egg	Zygote	Biosphere
draw	roots	head	embryo	rocks
gather	stems	organs	organs	fire
build	leaves	feet	body	air
program	flowers	wings	fetus	water
test	fruits	chick	human	earth
robot	seed	egg	zygote	biosphere

Table 5.2: Instructions of Creation

In table 5.2, the sets of instructions, conducted in such orders by the seed, the robot, human zygotes, chicken's eggs, and earth's atmosphere, are the procedures essential to generate new individuals of their own kinds.

The seed's vital instruction is to produce more seeds. A "series of preparations"—from rooting, stemming, flowering, to bearing fruits—are executed to produce new seeds. The egg's vital instruction is to produce more eggs. A "list of activities"—from assembling its head, organs, feet, wings to forming a chick— are delivered to produce more eggs. Humans also follow the same process of creation. A "work plan" to build a tool (an artificial creation), or a "family plan" to produce a child (a natural creation) is also needed.

Thus, in inscriptional creation, a sequence of instructions, a list of activities, a series of preparations, a work plan, a modular program, or a checklist is needed to accomplish a particular task, to cause something to exist, emerge, or be born.

When instructions are sequentially arranged together like in a list, the list of instructions becomes a procedure. For example, the four basic steps of building our model: from creating the ideas, drawing the design, gathering the materials, to assembling the project are specific instructions. When these four are combined in a series, they form a general procedure.

The four basic instructions are procedures in their own ways too. Each instruction is also a procedure because it has its own series of instructions as well, known as sub-instructions. The sub-instructions of each procedure are enumerated below:

1. Creating Ideas.
In creating an idea, the image of the object is pictured first in the mind. Such an idea is provided by something outside of oneself, like mother nature. The idea is extracted, borrowed, and

copied from her creations. They are inherent physical objects that came first before humans.

Thinking an idea and matching the idea with an object are two types of instructions. The instruction of thinking and the instruction of matching are sub-instructions which when combined form a specific procedure — the procedure of creating an idea.

2. Drawing the Design.
When the idea is matched with a physical object, it is codified by transferring or drawing it on a piece of paper. Sketching the image, assigning dimensions, specifying the details, and wrapping up the design are all pieces of instructions. These sub instructions when combined become another procedure — the procedure of drawing the idea.

3. Gathering Materials.
When materials, like wires, board, wheels, axles, and gears; and tools, like screwdrivers, pliers, and tweezers needed to construct the machine are gathered, another script is produced. These subscripts become another procedure — the procedure of gathering the materials.

4. Assembling the Project.

When all the materials are assembled while following a set of building instructions, the machine is now in existence. The sub-instructions of assembling, constructing, and forming the physical object is another procedure — the procedure of assembling a machine.

The tasks of creating the ideas, drawing the design, gathering the materials, and assembling the project are the general procedures which when combined in a queue, similar to a computer algorithm, becomes a process, the process to produce.

Codexation can also be illustrated through an investigative simulation known as the CAT experiment. This exploratory model was designed to examine 4 things: (1) How do bits of information bond together and transform into instruction; (2) How are procedures stored, embedded, and retrieved; (3) What triggers a procedure to turn on or off; and (4) How ideas originate from realities?

— THE CAT EXPERIMENT

In this simulation, a cat named Uno was used as a specimen. He was placed inside a sizable transparent tank of water and left to *sink for a couple of minutes*. After some uncomfortable wet ordeals, the feline was driven gently away out of the water to a waiting dry platform. The cat was thoroughly pat dried, warmed, and released afterwards.

From this experiment, the entire behaviors of Uno were simultaneously examined stage by stage, frame by frame, in slow motion, with the intention to investigate how instructions emerge into a procedure.

In the first stage of the simulation, when the cat was slowly submerging in the water while gasping out of air, his body managed to resurface through buoyancy and structural design. With no concept of survival, the cat endured all forms of discomfort by naturally kicking, swimming, and floating. His actions from sinking, bobbing, inhaling, exhaling, kicking, swimming, to floating were all individual instructions. When channeled on a queue, they eventually end up as a procedure — *a procedure tasked to stay afloat.*

In the second stage, the tired and depleted kitten once again submerged due to the small amount of aspirated water and heavy air in his lungs. His prolonged submersion caused shortness of breath that stirred a sensoric condition known as pain. The pain of suffocation (a switch) triggered the kitten to escape from the water by paddling towards a well familiar environment, a dry area. This second sequence of actions also shapes into a procedure — *a procedure tasked to escape*.

Thus, the summation to stay afloat and to escape, which was the procedure to get out of the water, was the main inscription of his whole ordeal. Getting out of the water was not engraved in his genes. It was an inlearned behavior that would never be forgotten by the kitten. It was not genetically inherited. If this terrifying incident happens the second time around in the future, the procedure of getting out of the water becomes second nature to the cat. It now becomes instinct, an inlearned instinct.

The pain of suffocation was the switch that caused the cat to escape out of the water and not bluntly because of the misconception that it must save its dear life. The task of saving one's life was not or never inborn or instinct. The idea of life or for this matter death was not even in his vocabulary. He survived the ordeal not because of intuition either. Instinct and Intuition were not inscribed in him too.

The countless events the kitten has endured in the water tank provide conclusive evidence that pieces of information are acquired from the environment (codified), transformed into a set of instructions, converged into a single algorithmic modular procedure, and imprinted as an inlearned instinct. This cumulative algorithmic task is called CAT; and, the self-logical linear sequential combination and rearrangement process of instructional tasks is called SCRIPT. The transformation of information into experience or behavior is also a form of codexation.

Other forms of codexation can also be illustrated through Dimetrix.

— DIMETRIX

Dimetrix, or Dimensional Metrics, abbreviated as Dx, is a specialized study that deals with space dimensions, spatial geometrical measurements, and linear fingerprinting. It is a branch of inscriptional physics that examines isometaspace closed structures called wylzan.

A point is merely an abstract concept. It has no length, width, or depth. A point is just a creation of one's mind. It does not even exist. It is not an object at all, but a mere representation. By intuition and observation, it can be interpreted by associating it with some distinguishing features. It can be described as a period at the end of this sentence. It can be named or tagged as "period or dot" to substantiate a physical form. When a point is labeled as a period, it exists now as an object. This example supports the idea of codexation, ideas become real.

Since the point is now in existence, it can transform into various forms, like for example a line. The line becomes a plane. The plane becomes a body of planes - a three-dimensional (3D). The body of planes becomes a body of bodies of planes (4D) and 4D becomes 5D and the dimensional progression goes on and on. This shows that a series of dots (material) is actually a series of instructions that can produce something to exist, such as a line, plane, or dimension.

Dimensions can also be codexated in many ways. In standard geometry, dimension zero is a point, dimension one is a line, a two dimension is a plane, and a three-dimensional figure is a cube. They can only be observed in a three-dimensional world. Although a fourth dimension is hardly imagined, its presence can be detected and measured by means of Photometrical, Originemological, Dimetrical, and Analogical techniques.

Dimensions can also be codexated by a set of spatial points or coordinates specified by variables. A three-dimension can be represented by the variables x, y, and z and written in the form (x, y, z) A two-dimensional as (x, y) and one dimension can either be an *x*, a *y*, or a *z* as in (x,0), (y,0), or (z,0). By following the sequence, a fourth dimension as (x, y, z, n) where "n" is another parameter such as time or a photon can be established. If "t" stands for time, the sequencing becomes (x,y,z,t). If "p" is for photon or light, the set becomes (x,y,z,p) or (x,y,z,l).

In a conceptual dimensional space, the position of a point can be visualized using a modified coordinate positioning system. This system forms a time frame that divides the past from the future. Think of it as a bridge between space and time, where time is another dimension and space is the other dimension. Although space evolves first before time, they are both continuously expanding since the birth of the first particle.

In isometaspace, the conventional X-Y-Z axes coordinate positioning system is used as a spatial basis for reference. These axes merge towards a common point of origin in a time frame (T-frame) called Crown. The coordinates are plotted into two-dimensional frames, intersecting evenly perpendicular to each center. The X-axis is pointing parallel within the T-frame; the Y-axis is pointing upward perpendicular to the X-axis, and the Z-axis is pointing away 90 degrees (or karats) from both X-axis and Y-axis. Their anti-pairs are also plotted on the backside of the Time frame. In order not to complicate matters, the front side of the time frame will be utilized. The T-frame, also called the Y-Z frame, is a chronograph spatial surface where lines are plotted.

Another way dimension can be codexated is through a coordinated fixated space using a grid positioning system (GPS). A grid system is made up of horizontal and vertical virtual frames, equally spaced and paralleled to each other. A sheet of graph paper is a good example of a grid. The graphing paper lines extend outward to form several virtual frames. These virtual frames are stacked like decks of cards. In an IsoMetaSpace, a dimension is a helical lateral iso-dimensional space divided into chronographic frames.

To visualize the dimensional concept, imagine you are on the other side of the globe on a windy summer day flying a colorful kite alongside the beach. Normally, to locate your exact position, your latitude (horizontal line) and longitude (vertical line) are extracted from a map. The map, which serves as a positioning tool, is a flat surface which is made up of latitudes and longitudes. Both are called variables and coordinated as a two-dimensional point. Meantime, to locate the exact position of the kite, it requires another kind of variable called altitude.

The location of the kite now lies in a three-dimensional point and thus your position is now in a three-dimensional space as well. This happens because the latitude is ground zero.

Moreover, another variable exists between you and the kite. Time is the other variable. Much like latitude, longitude, and altitude, time can be perceived and measured. When two intersecting dimensional frames orient a three-dimensional space, time becomes the third frame. How is this so?

Let us say we want to create a cube (1" x 1" x 1") in a two-dimensional frame. To do this, we plot this cube on an X-Y and X-Z frame graph. In the X-Y frame, there is one unit plotted on the x-axis and one unit plotted on the y-axis. In the X-Z frame, there is one unit plotted on the x-axis and one unit on the z-axis. If we intersect these frames in a 90° angle, wherein the x-axes merge into one, and shuffle these frames so that the x-y frames go to the right and the x-z frames extend upward, the configuration creates a virtual isometric cube.

In the above example, an X-Y frame and an X-Z frame - two frames intersecting each other to form a 3-dimensional figure are created. Within this configuration, a third frame exists, the Y-Z frame. Y-Z frames are the graphical representations of

Time. These time frames or chronograph frames are plotted in such a way that the Y-axis represents a 3600^ karat revolution and the Z-axis divides into 365 days.

Dimensions can also be codexated using representational models. For example, a single card drawn out from a deck of cards when viewed on its edge provides a field of what a one dimension is. The edge is simply a line. Flip the card in a standing position facing you, it indicates a two-dimensional space with four edges called the perimeter. The point of origin is now two-point lines. Extending the card backward by placing the remaining cards in the deck behind the card yields a rectangular shape. The structure creates a three-dimensional space with twelve edges and three-point lines. The point of origin evolves from one to two to three and four-point lines. The fourth dimension is made up of four-point lines shaped like a roller deck of cards in the form of a donut. The mirror image dimension is symmetrically opposite, creating two new worlds of space - the inside hole and the outside space - much like the shape of a doughnut or a torus.

Dimension, in a simpler version, can also be codexated using a typical ordinary house. A house is usually made up of several rooms: a bedroom, a kitchen, a garage, a bathroom, a backyard, and a front yard. These rooms have their own original universe or characteristics. Like in the bedroom, objects such as the bed, drawers, and lampshades are usually found. In the kitchen, ovens, fridges, dishwashers, sinks, and cabinets are present. In the living room, a sectional sofa, a bar, a futon, and an entertainment center for flat-screen TV and gaming are displayed. The compositions of each room are exceptionally unique. Each dimension has its own individual trademark. When they are gathered all together, they form a bigger dimension - the overall space of the house.

All the rooms are now inter-dimensionally connected in one roof of a three-dimensional "house". If someone sitting comfortably in the living room is watching television, such a person is aware that there are other rooms in the house. But, because he is staying in the living room, he obviously doesn't see the other rooms or the other three-dimensional spaces in the house. A fine line separates and connects each room with the other room. In Isodimesional Morphical Space Figures (IMSF), invisible residual lines that interlink or interconnect dimensions are accessible. They are lines that separate every dimension. These dimensional lines are the doors that lead to the House of Multiverse. Through linear fingerprinting, multi-dimensions can be detected.

On the other hand, linear fingerprinting can also be viewed in simpler ways. As we know, everything has density, just like all things have mass and volume. Fluids like water, oil, honey, and alcohol, when poured in a graduated cylindrical slowly one drop at a time, can create multiple layers of dimensions separated by colorful residual lines. These layers, rooms with individual characters, provide clues that dimensions can be oriented in vertical multi-layers as well.

The sea, land, atmosphere, and outer space are examples of dimensions too. The sea is a dimension with its own characteristics. The atmosphere is another dimension with its own surroundings. Under our feet is another dimension that comes with its own characteristics. And above us, outer space has its own conditions totally different from the three as well. All of them are separated by dimensional lines.

If linear fingerprinting can be the next thing that might send us to the next wormhole, time travel, or beam me up scott, can a three dimensional entity be able to perceive someone who lives in a four, five, or nth-dimensional space? If an alien lives in one of these dimensions, can humans be able to see this space being? (The Dimetrix Argument).

In summary, codexation can be in multi-forms. It can transconverse from ideas to realities, from instructions to materials, from experience to behaviors, from abstracts to physicals, from equations to designs, or from space to shape. Whatever it may be, all these forms have one thing in common, they have the capacity to bring something into existence.

CHAPTER 6 — INSCRIPTION BY DESIGN

"Inscription is an inherent property of matter."

INscription by Design (ID) is a theory that claims everything is an intuitive object inscribed with an embedded set of inherent instructions. This set of instructions is known as Inscription (Lawsin 1988).

— EMBEDDED INSCRIPTION

Inscription is an inherent property of matter just like mass, weight, density, and volume. It is a set of instructions embedded in every object. Plants, animals, geometrical shapes, materials, by-materials, non-materials, and everything else are embedded with inscription.

Dots are examples of objects made up of inscriptions. They are in essence individual instruction in forms. When they are aligned side by side in a row, their inscription is to form a line. The dots are the set of instructions that created the line.

Lines have their own inherent instructions as well. When four lines are connected corner to corner in any given angular degree, they either form a square, a rectangle, a rhombus, or a parallelogram. These geometrical figures emerge because inscriptions are embedded in their structural designs.

A circle is another example of an object that consists of inscriptions. It is made up of instructions in forms of each part such as its circumference, diameter, and radius. The circumference is the line that shapes the perimeter or boundary of the circle. The diameter is the line that goes halfway through the middle of the circle. And the radius is the line half of its diameter. These lines are the basic set of instructions that shape the circle. Through these lines, the circle emerges. Without inscription, the circle will never exist at all.

Its set of instructions also generates an inherent inscription in the form of an equation. It always exists whether the circle is big, small, or in-between sizes. The equation is always inherently embedded side by side with the circle.

The circumference(C) is the most visible part of the circle. It shapes the perimeter of the circle. The perimeter is the product of three instructions: radius(r), pi(π), and a constant (2). These embedded inscriptions can be converted into the equation, such as, $C=2\pi r$. The formula tells us that whether the values of the circumference and radius go big or small, the pi(π) will always be equal to 3.1416. The proofs are enumerated in table 5.1.

Circles	Circumference (C)	Radius (r)	Pi (π)
Big	125.6	20	3.14
Medium	94.2	15	3.14
Small	62.8	10	3.14

Table 5.1

Instructions also manifest as social individuals such as in the society of ants. Ants are organized social groups that are made up of individuals. The community is driven by a set of instructions represented by its members. Finding food, cleaning the nest, and protecting the queen are independent instructions. When accumulated as a list, they emerge as inscription. This embedded inscription runs the system of the whole community of ants.

Inscription is also substantiated in formula progression. In two-dimension, the formula for a square is L x L = L^2 where L stands for length. In three-dimension, the formula becomes L x L x L = L^3. In four-dimension, the formula can be L x L x L x L = L^4. In nth-dimension, the general formula is L x L x L ... x L = L^n. Thus, inscription can also be in the form of dimensions.

The formula for a circle, in two-dimension, is $x^2 + y^2 = 1$. In three-dimension, the formula becomes $X^2 + Y^2 + X^2 = 1$. In four-dimension, the formula is $X^2 + Y^2 + X^2 + W^2 = 1$. And the sequence can go on and on as it produces randomized multi-dimensional shapes.

In these progressions, the variables are actually individual instructions which when combined in series form an inscription. Thus, inscription can also be in the form of equations.

— INTUITIVE OBJECTS

Everything is an intuitive object. Objects are intuitive because they carry instructions of their own capable of doing self-calculations without knowing (aneural) them. This instructional ownership is the first Law of Inscription by Design.

Whether simple or complex, objects are naturally made up of inscription. These inscriptions are extracted from its own structural configurations, shapes, or designs. When triggered by an outside force, their inscriptions activate and generate the phenomenon known as Animated Reality.

For example, an autognorics prototype named KIM, which stands for Kinesthetic Intuitive Machine, is a complex biomechatronic spider. Its parts like the gears, axles, beams, motor, and battery are individual instructions. When energized, this set of instructions animate KIM via its structural design and configurations.

KIM is animated due to its embedded inscriptions and its material designs. It walks like a spider, moves like a spider, stares like a spider, jumps like a spider, and behaves like a spider because it is designed like a spider.

KIM's legs, which are merely a bunch of lines (levers), can stride, bend, crawl, stand, jump, and play dead because of configuration and designs. When these legs are replaced with wheels (circles), KIM's walking behaviors will definitely change. Instead of walking, KIM will either rotate or roll like an automobile. KIM behaves this way because wheels have their own inborn instructions as well. When the design is changed, so are the instructions. This structural change is the second Law of Inscription by Design.

Inscription defines the importance of designs in all living and nonliving things. Snake slithers, crocodile crawls, kangaroo hops, tick jumps, plant trails, shark swims, fungi spores, bacteria propel, and stones rolls because of their unique physical forms. The design of an object dictates the set of instructions it owns or carries..

Inscription by Design "pens" the script of everything. Whether humans, bacteria, or a piece of wood, everything is programmed through its structures, configurations, and designs. This principle is the third Law of Inscription by Design.

KIM, being a complex machine, is also a complex system. She is made up of smaller systems and smaller sets of instructions. When parts are added to the machine, the rules of the game so to speak also change. When new parts are added, additional instructions are added too to the inscription. Thus, when Material increases, Instruction increases as well. This next principle is the fourth Law of Inscription by Design.

Furthermore, as new parts are added, the inscription becomes more intuitive, the materials become larger, and the supply of energy gets bigger. However, the more complex the system is, the more resources are needed. The more complex the system, the more complex is the design or configuration too. This is the fifth Law of Inscription by Design.

Alongside the script of acquiring information, emerging as instructions, converging into a modular list, compiling into a cumulative algorithmic procedure, and converting into law, Inscriptions also transform logically into Intuition.

Intuition evolves naturally in four main instructions: the Accumulation of Information, Compilation to Instruction, Translation to Procedures, and Inscription to Law. Intuition, sometimes called Inscriptional Logic, is an invention of Inscription by Design.

Just as babies and dogs get information from their surroundings, machines also acquire information from their environments. Information is the key factor in creating a machine that is living, alive, and with life.

When information is received as input, the same information is sent out as output. The input that zips into the system is equal to the output that zips out the system. What comes in, must come out. When water flows in a garden hose, it will come out as water. (water=water). A marble that rolls inside a plastic tube will exit at the other end of the tube as the same original marble (marble = marble). When the word "hello" is spoken on a telephone, the same word "hello" (olleh) will come out as an output (hello=hello). All these examples demonstrate the principle known as the zizo effect.

However, there is one experiment in physics that defies this principle. Or maybe not! A cardboard, with a pinhole at its center, is used to project a physical candle on a wall. The candle, which serves as an input, is projected through the hole as an output, and comes out itself as an inverted image. Interestingly, the physical candle did not appear as a physical candle but an abstract candle. Since the output of the candle is not equal to the input, does this mean that the physical candle is just an optical illusion? Since an abstract image of the candle is projected, does this mean that the physical candle is all nothing but an illusion? If so, then, can we say that the material world we live in is after all not really real but actually abstract in nature. If this supposition is true, then the hypothesis that everything doesn't exist is therefore correct.

To explore this setup further, let us reexamine the four examples and inspect if there are discrepancies with our models and reasonings. In the first three examples, (the three objects: water, marble, word), if we shrink the garden hose, the cylindrical tube, and the string with the same thickness as the cardboard, the marble can come in and go out as the same marble, the water can come in and go out as the same water, and the sound can come in and go out as the same sound.

Following the same examples, the physical candle should also come in and go out as the same physical candle. But why did the image become abstract instead? Where did the image of the candle on the wall come from? Well, the answer lies in the power of geometry and equations.

— THE CUE

When an output is zipped out from a system and sent to another system, it becomes an input in the second system. The output signal becomes a trigger mechanism that switches on and off the second system. This switching signal is called the Cue.

Like in a domino run, every standing tile is a switch. When a cue, an outside force, taps the first domino, it switches on its internal instruction and taps the second domino. The internal cue of the second domino is switched on and becomes the external source of the next domino. The third domino now becomes another external cue to the next domino and the whole process of this external and internal switching goes on and on until the last domino is toppled. Once the cue reaches the last domino, it is dissipated freely to the environment. What is interesting about the whole domino effect is that when the external cue is consumed by another system, the internal cue of that system is engaged to activate another system. The external cue always generates from another outside source.

When this never-ending event goes on and on where the cue is consumed and replaced, the occurrence becomes a perpetual recursion that is not created or destroyed but purely cycles around. This phenomenon of perpetual motion in real time is always transpiring in both micro and macro scale!

Eventually, inscription becomes mechanical when performed repeatedly. The series of instructions sequentially creates behavioral experiences while its material design creates memory functions. This principle of intelligence is the General Law of Inscription by Design. Because of instructions and materials, the intelligence of creation evolves.

— ANIMATION BY DESIGN

According to ID, every material always comes with embedded instruction; and every embedded instruction comes with intuitive material. When the two collectively work as a system, they mutually operate and transcribe into sets of behaviors. When they are energized, they generate the emergence of animation known as Animation by Design.

In the picture, the emerging behaviors generated by the machine when energized are naturally caused by its intuitive materials and embedded instructions. The material parts that shape the machine are the pieces of instructions. Its legs or linkages are made up of four instructions embedded in each physical form. When these legs are in sequential motions, like walking from left to right, left to right, they establish a set of instructions known as a sequential procedure. The legs, which are levers in nature, behave the way they are because of their designs and structures. If one of its legs is removed, the object will behave totally differently from its original behavior. Technically, the behavior changes because an instruction from its original inscription (program) is removed or deleted.

The rectangular shape, that holds its body and parts, is also another type of instruction. Its job is to serve as a receptacle. Its eyes are instructions too. Their task is to detect distances in varying lengths. Its microcontroller is a compendium of instructions as well. It serves as a gateway of inputs and outputs that animates the model. The battery is also an instruction. Its task is to provide energy that makes the object alive.

The object is alive because it is energized by a self-rechargeable battery and animated by its own set of instructions through its structures, compositions, configurations, & designs. Being alive doesn't mean it is living or with life.

And, when pieces of information are transformed into a set of instructions, the set emerges as Inscription. When inscription is replicated to produce similar objects, it becomes a universal law. When the emergence of information subsequently creates the intelligence of creation, everything has the natural tendency to Coexist, Reorganize, Expand, Apair, Transform, Empower, and Select — (to) CREATES.

CHAPTER 7 ALIVE, LIVING, LIFE

"Life emerges only when it is alive and living."

WHEN I was in high school, my biology teacher told us that for something to be living or with life, the object must have or had the following specific characteristics:

— CRITERIA OF LIFE
- It consumes food as energy.
- It is moving or in motion.
- It reproduces with an exact copy of itself.
- It reacts to its surroundings.
- It is made up of cells.

However, the last criterion seems somewhat shaky in the sense there are so many non-cellular microorganisms that exist without cells and are alive. There are also other living organisms that lack one or two of the criteria, but are considered alive, living, or with life.

Besides, there are other features that can be found in some living things like walking, talking, flying, seeing, thinking, swimming, and feeling that are not found in other living organisms. Plants and trees don't have these characteristics, but this doesn't mean that they are not living or alive.

On top of this, there are non-living things, such as robots and space probes, that also hold the same comprehensive criteria of life. These man-made objects can talk, walk, see, feel, think, eat, and even die. They even exhibit mechanical emotions, machine learning, and aneural consciousness. They act and interact with their environment. They consume energy, in motion, and program to reproduce. Some even have mechanical organs similar to our biological organs and sensors. With all of these similarities, what makes an object a living thing?

Medical scholars and legal experts, on the other hand, have different views about life. They define life according to the legal definitions of death. To these individuals, death means:

- Total failure of the heart.
- Total failure of the lungs.
- Total failure of the brain stem.

If we characterize life by how medical experts define death, then an object is alive if it has a functional heart, lungs, and brain. But obviously, the definition does not apply to all living things as well like in the plant kingdom. Trees and flowers do not have hearts, lungs, or brains; yet they have life.

The Monera is a living organism, but it does not have organs. This life-form walks without feet, eats without mouth, digests without a stomach, and reproduces without reproductive organs. Octopuses, cuttlefish, nautiluses, and squids all have three hearts that pump blue blood and change skin colors faster than a chameleon, but still they lack some life major organs and yet they are considered living things. These examples suggest then that the heart, lungs, and brain serve no purpose at all in the definition of being alive or with life. The criteria of life, therefore, varies with different creatures.

In comparison, if we characterize death parallel to how science defines life, then we can consider an object dead when it is no longer moving, consuming energy, reproducing, and reacting with its environment. This latter definition seems more reasonable since these traits can be generally applied to both living and nonliving things.

Thus, if we arrange the criteria based on their levels of importance and reduce them to one category through the process of elimination, then energy is the only valid ultimate criterion that is left.

Natural objects, either living or nonliving, cannot be in motion without energy, reproduce cells without energy, and react to their surroundings without consuming energy. Energy produces force, which makes objects alive. It creates motion through its force. It generates signals. It animates things. It flows from one system to another system, from one family to another new family, from the physical world to the abstract world. It is boundless and continuous. It makes both living and nonliving things alive. Energy is the primary criterion that warrants the existence of being alive, living, or with life.

In Autognorics, being alive, living, and with life are three different items much like consciousness is different from awareness. Their differences are individually investigated here as we recall the principles behind the following literatures: (i) The Caveman in the Box Trilogy (origin of information), (ii) The Human Mental Handicap (acquisition of information), and (iii) Inscription by Design (algorithm of information).

Aside from these, we also investigate consciousness, a term believed to be synonymous with life. We also explore the levels of consciousness based on the research of Philippe Rochat on his thesis the "Five Levels of self-awareness as they unfold early in life", the stages of consciousness based on medical research, the hierarchy of consciousness based on evolution, and the seven orders of consciousness based on the mechanics of creating a living machine.

And finally, we take a look at how humans become a living being, and through the new seven criteria of life, we redefine the meanings of being alive, living, and with life.

(i) In the Caveman in the Box Trilogy, we learned that Mother Nature is the source of all information. She is the giver, database, brain, and keeper of information. We also learned that Information is gained only in two ways: by choice or by chance. It flows one-way that starts from nature to the mind It transforms into instructions, procedures, and the emergence of behavioral experiences. It is sensed, copied, matched, lived, and experienced. Information makes who we are today.

(ii) In the Human Mental Handicap, we learned that consciousness is basically the ability of an object to match things with other things. The concept is based on the Codexation Dilemma that proposes: "No human can think of an abstract idea without expressing it outside the mind by matching it with a physical object". This irrefutable assertion was originally discovered from the isolation experiment that affirms information is inherently acquired from nature and flows into the mind (scriptional jump), an irreversible process. This suggests that before one can think of something, a physical object must be present first. This one-to-one correspondence is the litmus test or key indicator of consciousness.

(iii) From Inscription by Design, we learned that everything is an intuitive object that owns some inherent instructions in forms of equations or laws. This series of instructions produces unique behaviors of its own when cues are applied on its design. Through Inscription by Design, objects can store and process information without the need of the brain.

— CONSCIOUSNESS

The brain is believed to be the seat where consciousness resides. Humans are conscious because evolution furnished them with brains. However, this idea is comprehensively refuted here because of the following reasons:

First, there are some animals or organisms that are conscious but lack the physical brains. Examples of these aneural creatures are plants, starfish, jellyfish, sea sponges, corals, bacteria, fungi, sponges, sea urchins, sea squirts, sea cucumbers, anemones, archaea, protists, and hydras to name a few.

Second, plants and trees are conscious because they interact with each other. They process information just like all other animals. They can reproduce, react with their environment, protect each other from intruders, adapt, and metabolize. They eat, sleep, breathe, grow, and die just like all other living things. They are aware because they are equipped with sensors. Plants and trees are conscious because they have the abilities to choose options and not because they have brains.

Third, consciousness can be properly defined without the need of mental reasoning or with the help of the brain. The reasons are as follows:
1) Consciousness is about association.
   1.1 Mimicking or copying
   1.2. Matching or pairing
   1.3. Discovering or inventing
   1.4. Inlearning or acquiring
   1.5. Parroting or aping

This one-to-one correspondence is known as Associative or Correlative Consciousness (Definition-1, Lawsin 1988).

Consciousness is the ability to match things. This ability to associate things is the crucial indicator of consciousness. A baby can randomly play bricks of legos and stack them at various heights and shapes at later infancy without even knowing about stacking, legos, heights, and shapes. Even birds can equate nests for babies, worms for food, fly for predators. Their behaviors to match or copy things from their parents are indicators they are conscious. The experiential interactions of plants, corals, microbes with their surroundings apparently also suggest consciousness. This ability to match or pair things without mental intervention is known as Aneural Consciousness.

2) Consciousness is about the equations — If x is conscious with y, then x is conscious. If x is alone, surrounded by nothing, then x will never be conscious.

In this definition, consciousness is based on two basic variables: X and Y. If one of these two elements is missing, then consciousness will never emerge. To be conscious one must be aware of oneself and one's surroundings, where in this case, the surrounding is another person. Thus, to become conscious, two things must be present: a being and a surrounding, or a being and another being.

Although consciousness is believed to be synonymous with thinking, as interpreted by the famous quote "I think, therefore I am" by Rene Descartes, and is believed to be associated with thoughts, existence, and the mind, these ideas are all completely incorrect. Rene has forgotten that there are beings without brains that exist and are conscious, and there are beings that are not conscious but have brains. For example, children and infants, in general, are not conscious and self-conscious at a certain early age. Thus, to be conscious, one neither needs to think nor to have a brain. One simply needs to match or choose. Thus, to rewrite Rene's quote about consciousness in simplified form measured according to the Lawsin dictum: "If I can match x with y, therefore I'm conscious." (Definition-2, Lawsin 1988).

3) Consciousness is relative to the inlearned behaviors or traits of the organism:
3.1. Species with babies is a conscious being.
3.2. Species that live in houses, caves, nests, undergrounds, are conscious beings.
3.3. Species that sleep are conscious or once conscious beings.

3.4. Species that recognize objects are conscious beings.

3.5. Species that defend themselves are conscious beings.

3.6. Species that mate are conscious beings.

All of these inlearned behaviors are signatures of consciousness. Inlearned behaviors are the Collaborative Determinants of Consciousness (Definition-3, Lawsin 1988).

4) Consciousness is an emerging process that runs in an orderly inscription:

4.1. Information transcodifies to physicals,

4.2. Physicals transform into actions or movements,

4.3. Motions transfer into repetitive, or animated actions; and,

4.4. Inscripted animations translate into the persona of being.

This emergence is called the Grand Script of Consciousness: (Definition-4, Lawsin 1988).

5) Consciousness is both physicals and abstracts:

Consciousness is a two-step action. It revolves between physicalness and abstractness. It is classified into four categories, namely: (i) Physicals to Physicals, (ii) Physicals to Abstracts, (iii) Abstracts to Physicals, and (iv) Abstracts to Abstracts. When an idea is codified with another idea, like dreams, it is abstracts to abstracts. When an idea is codified with an object or object with an idea, like inventions, it is abstracts to physicals or physicals to abstracts. When an object is codified with an object, like play, the transensation is Physicals to Physicals. Dreams, inventions, and play are solid computational measures of Aneural Consciousness. The process of pairing physicals and abstracts is known as Codification or Associative Codexation.

The ability to interact one to one — by association, representation, tags, labels, signals, codes, symbols, and the primitive way of pointing — is the basic element that measures consciousness. Codexation is the primary indicator of consciousness. (Definition-5, Lawsin 1988).

6) Consciousness is a product of generated emergence: According to codexation, matter is made up of two parts: Materials and By-materials. In the macro world, things with mass are classed as materials. Things generated by materials are classed as by-materials. By-materials are by-products of the materials. By-materials are parameters invisible in nature (which the Big Bang Theory has failed to acknowledge). Things like apples, rocks, air, water, and fire are examples of materials. Things like temperature, pressure, gravity, density, heat, and

electricity are examples of by-materials. By-materials only emerge when their elemental materials are present.

When both Materials and By-materials interact and evolve into a complex system, parallel to the gears and dynamics inside a clock, they become energized, automated, animated, and codified. This elaborated emergence of codexation infuses the phenomenon of mechanical consciousness. This type of emergence is known as Generated Consciousness (Definition-6, Lawsin 1988).

7) Consciousness is Information Materialization.
Information Materialization is the ability of an organism to translate physical objects to abstract ideas. Technically, by definition, I.M. is the transcodification of the physicals (outside the mind) to abstracts (inside the mind). Without the inherent world, translated collectively as information, consciousness will not exist. Consciousness exists because materials and information exist.

When objects are transcripted into ideas, the transcription is codified. This transcription or scriptional jump from materials to ideas is known as Materialization of Information (Definition-7, Lawsin 1988).

In neuropsychology, consciousness is viewed in terms of alertness or responsiveness characterized by the patterns of electrical activities in the brain recorded by a device known as electroencephalograph. Psychoanalysts, meanwhile, define subconsciousness as a mental process occurring just below the level of awareness. It is a zone between the conscious and the unconscious. Although in popular usage, subconscious and unconscious are often taken as synonymous, the distinction can be defined by examples like sleeping is to subconsciousness as comatose is to unconsciousness.

Here are the levels of consciousness according to Rosalt:
0. Self-obliviousness - absence of self-awareness
1. Confusion of extended dimension
2. Differentiation of two worlds
3. Situational exploration
4. Identification of "me" - a cognitive index
5. Permanence - first person experience
6. Self-awareness - third person experience

Next are the stages of consciousness according to medicine:
1. Comatose - less conscious
2. Anesthesia
3. Deep sleep

4. Inhibitors

5. Sleep walking

6. Epilepsy

7. Wakefulness - more conscious

And the levels of consciousness according to evolution:

1. Humans

2. Orangutans

3. Trees

4. Corals

5. Bacteria

6. Weather

7. Universe

When the first man on earth discovered softness and hardness while using objects such as stones and water, not knowing all these objects in terms of mental understanding, he gave birth to consciousness.

Consciousness is not a process in the brain. It is one of the evolutionary stages of life that emerges when an object uses energy, reacts with the environment using its sensors, and associates things with other things.

Since then, the nature of consciousness has been debated for hundreds of centuries by scholars. Today, the ancient questions of what, how, and why of consciousness, will now be put to rest as we acknowledge them based on the seven stages of life.

(1) The Functional Question: Why is there consciousness? In simple words, there is consciousness because of inscription by design and generated interim emergence.

Philosophically, the question is definitely ambiguous. This is like asking why dogs exist or why do humans exist? The inquiry is open to over one interpretation where the varying opinions could either be accepted as right or wrong. Originemologically, consciousness exists because of sequential instructions that give rise to logical experiences; and the structural designs of the object that produce functional mechanical behaviors.

(2) The Explanatory Question: How does consciousness exist? Consciousness is an emergent that exists when the subject is alive, equipped with sensors, and able to codify.

Primarily, consciousness doesn't need a brain to exist. It simply needs a subject that uses energy and is equipped with a sensor, a biotic receptor that captures information just like a magnifying lens (sensor) that detects or projects an object (physical) with a matching reverse image (abstracts). The ability of the lens to transense physicals to abstracts is a tangible example of consciousness. Consciousness is not a product of the brain but a process that doesn't need the brain.

(3) The Descriptive Question: What is the purpose of consciousness? It emerges to produce intuitiveness, inlearness, livingness, and lifeness. These last four orders are the elements that spark the phenomenon we humans described as LIFE.

Notice that Life emerges when one is alive and living. Being alive, living, and with life are three different things according to the seven orders of life: namely, Aliveness, Awareness, Consciousness, Intuitiveness, Inlearness, Livingness, and Lifeness.

A more comprehensive discussion of these orders will be discussed in the next chapters while we are building a living thinking machine. Inspite of this, let me show here first some more fascinating thought-provoking notions about the real essence of life that I have discovered in my experiments.

— THE PARADIGM SHIFT

(1) The Codexation Dilemma — No human can think of something without associating his thought with something else.

(2) The Science Quandary — No human can reproduce an exact duplicate of Nature's creations.

(3) The Generated Interims — Things exist because other things cause them to exist.

(4) The Unicorn Belief System — Humans are governed by Natural Laws and not by belief systems.

(5) The Algorithm of Queue — Humans are animated inscripted intuitive entities. They are mechanical, scriptional, and casual autognorics.

(6) The Information Codification — Human's ideas cannot be codified into physical realities unless an external material inherent world like Nature is present.

(7) The Law of the Second Option — Humans, like everything else, are always governed to do only two things at one time, either This or That.

(8) The Circle of Perspective — No two humans can perceive the same object exactly the same at the same time.

(9) The Irreversible Jump — Humans originally acquire information from Nature.

(10) The Guesswork Predicament — All humans' ideas are replaceable and therefore are all circumstantial, incidental, conditional, or guesswork.

(11) The Bandwagon Effect — All humans' language system, social system, mathematical system, the money system, the calendar system, and all other systems created by mankind are mandated through the Power of the Standard. The Standard is the Rules of the Land which are generally agreed upon through the power of the Majority.

(12) The Birthday Riddle — All humans celebrate their birthdays not only once in a year, but twenty four times in a year.

(13) The Belief System Conjecture — All humans' beliefs are influenced by their environment and culture and therefore are circumstantial or merely guesswork.

(14) The Black Train Effect — Everything humans perceive, or sense is all but illusions. Nothing is real.

(15) God's Identity Crisis — Humans created God. We borrowed the idea from nature through imagination. God exists only in the mind, a mental impression, a concept, a notion that will never ever be codexated since god is not inherent.

(16) The Creator Argument — the universe can be created without the intervention of a God or creator. The twoness of both particle and instruction causes things to emerge via

inscription by design. The evolution of instructions designed the intelligence of creation.

(17) Cosmogenical Argument — If God created the earth, then the various other planets, suns, and galaxies that came first before our milky way were created by someone else, who.

(18) Originemological Argument —Everything has a beginning. If God exists, therefore, he has a beginning. If god exists, then he entered into a process. If god exists, then materials and instructions existed ahead of him first before becoming a being.

(19) Phrenological Argument — If God created the mind, but the mind cannot detect truth and reality, then God will never ever be known.

(20) The Heaven Argument — also known as God's Boring Argument, claims that living in heaven or in hell is a dull, boring, tedious, monotonous life that lacks variety, interests, and excitement for the rest of someone's existence.

(21) The Big Bang Theory Argument — materials and its by-materials must coexist to give birth to the universe. Without the by-materials, such as mass, temperature, pressure, volume, density, inscriptions, the Big Bang will never ever take place.

(22) The Duality of One Paradox — Everything is both living and dead, in motion and at rest, real and not real, existent and non-existent. Birth is the beginning of Death, the final genetic code of life.

(23) Astral Projection — the soul is a hallucinatory emergent generated by a "dying" or disoriented mind, or a by-product caused by irregular or abnormal environmental and bodily conditions. When the physical body and mind die, the soul dies as well.

(24) The Lawsin Syndrome — Humans reality can be altered or distorted due to medical conditions, health issues, or cognitive bias. A condition I have experienced many times as I watched my face, half solid and half empty.

(25) Shape and Space — Everything is made up of invisible shapes and empty spaces. Since atoms are not alive and humans are made up of atoms, then we have never been alive.

(26) The White Light — Without light, everything is black. Rainbows will never emerge. Colors will never exist and be known on earth.

(27) Tricks of the Senses — Without sensors, awareness and perceptions will never exist. Life will never be known.

(28) Present Past — Everything is always in the past.

(29) Glass Synchronicity — Everything can be both moving and not moving at the same place and at the same time.

(30) Hello-olleH Reversibility — all things humans hear and speak, see and think, perceive and sense are always in reverse order.

CHAPTER 8   SEVEN ORDERS OF LIFE

"Life is an evolutionary process."

IN the last chapter, we learned that objects have life if they have the following seven criteria: they consume food as energy (eat), take and expel gas (breathe), move or in motion (perform), reproduce an exact copy of themselves (replicate), grow with their surroundings or environment (thrive), respond with their sensors (sense), and are made up of cells. Aside from these, some of them can talk, walk, see, feel, think, swim, fly, climb, regenerate, and die.

However, it turns out that there are also objects that lack one or more of these criteria but are alive or living. For example, the non-cellular microorganisms that exist without cells are living objects. A seed, a non-living thing, that turns into a tree and becomes a living thing. A virus, a chemical machine, that becomes alive when it lives with a host. A neuron, a non-living thing, that becomes a living thing when it works with an organic network.

Also, being alive can be interpreted based on the criteria of being dead. To be dead means the heart, lungs, and brain stem are all totally not functioning anymore.

But of course, not all living things have brains, lungs, or hearts. Trees, flowers, and jellyfish are living things, but they do not have hearts, lungs or even brains. The Monera, a living organism without organs, can walk without feet, eat without a mouth, digest without a stomach, and reproduce without reproductive organs. They are all living things.

It seems there is no definite criterion that's designated to identify when an object is living or alive. However, by comparing all the old criteria of life through elimination, there is one common vital factor that shines among them. For something to move, reproduce, react, and make its heart, lungs, and brain function, it needs one important thing – Energy. Without energy, life is nonviable or will never exist.

Notice that the old way of characterizing a living object is somewhat vague or too shallow. There is no definite fixed foundation to describe when an object is alive, living, or with life. Because of this issue, while performing various exploratory experiments in making a mechanical life-form for several years, I discovered that life emerges in seven stages. Life is an evolutionary process. It evolves from being alive, living, and with life. Life is governed by Seven Orders, namely:

1. Aliveness
2. Awareness
3. Consciousness
4. Intuitiveness
5. Inlearness
6. Livingness
7. Lifeness

— ALIVENESS

Aliveness or Self-Animated State of Being:

Aliveness is the first order of life. An object is alive when it uses a continuous supply of energy to animate itself. Its energy is provided by an outside source. Energization is the first basic indicator that identifies when something is alive.

When an object uses a certain amount of energy, the energy produces force. The force activates the movements, actions, and behaviors of the object. The animated movements are automated mechanically which are carried out by the object body's shape much like the mechanical behaviors of why a person walks, crocodile crawls, snake slides, and car rolls. All the animated structural actions and behaviors are not shaped by accident but caused gradually by the dynamics or mechanisms of one's body's designs, materials, constructions, configurations, structures, and instructions through a natural phenomenon known as Inscription by Design (I.D.).

The self-mechanical automation emerges because of the force or series of force provided by the source of energy. The force activates the instructions in the object and brings it to aliveness. A force is a push or pull that turns on and off a system like flipping a switch On and Off. When a force is given continuously by an energy source, it is called impulse. When a force is given continuously at a distance, it is called work.

Force and Energy are two different things. Energy provides force. Force activates a system. A system, which is made up of instructions, becomes alive and animated when its instructions are activated continuously while supplying it constantly with a certain amount of force from an energy source.

For example, a jellyfish, an animal without a brain, heart, or blood, can energize itself while consuming energy aimlessly in the sea without even knowing it since it doesn't have a brain. Although it is brainless, it is considered alive because it can energize itself by ingesting food.

A newborn baby is typically considered alive when he consumes energy on his own. But technically, the baby at this age is not yet alive because he can't self-recharge or self-energize without the help from external sources. The energy usually comes from his mother's milk, an external source.

The states of being exhibited by the baby and jellyfish are categorized as Aliveness or being Alive. The energy is provided by external sources: the baby's milk from his mother and the jellyfish's food from the ocean. In both cases, the subjects are technically considered alive since energy is used individually. Aliveness or being alive emerges when energy is used.

Orchestrated or Programmed State of Being:
Another form of Aliveness is known as Programmed Animation. In my Youtube channel, I posted a video of a machine that I have been working on in many of my

experiments. The mechatronic is alive because it self-consumes energy by simply docking on a charging station. Its ability to self-dock when its energy is low is an indicator of aliveness. It is alive because it uses energy. The automatic actions of going back and forth to its docking station seem to be a sign of awareness or consciousness, but its behaviors are actually manipulated by a series of instructions extensively coded using a computer program. The computer code is embedded in its microcontroller (a configuration of switches) while the mechanical inscription (a series of instructions) is embedded in its structural design. The animation emerges because of the mechanical inscripted instructions of its structural design. This type of aliveness is called Orchestrated or Programmed Animation.

— AWARENESS

The Sensoric or Aware State of Being:

The second order of life is Sensory Awareness. In this state, the object is alive and aware. To produce awareness, an object needs two requirements: (1) it uses energy; and (2) it is equipped with sensors. The intuitive sensor is the key indicator of awareness. Through the sensors, objects can gather information from their environments. Awareness is the ability to sense or perceive things through sensors.

Intuitive sensors are technically called Exyzforms. They are pre-designed much like the eyes for seeing, ears for hearing, or neurons for thinking. Without sensors, objects are unable to react to changes in their environments or make decisions on how to react or behave in a situation. Sensors are naturally or artificially pre-designed with inherent instructions.

When a newborn leaves a womb, the baby's first reaction is usually to cry. The behavior is not a sign of consciousness but merely a sensoric awareness naturally triggered by the temperature, odd odors, and loud noises unfamiliar to the baby. The reaction is carried out by his biological sensors, such as the ears, nose, and skin. The reaction is simply a kind of mechanically triggered sensation. The sensation transpires because the subject is equipped with intuitive sensors and not by the false belief of owning a brain. His biotic sensors are naturally designed to sense things mechanically or inscriptionally. These mechanical sensors make the newborn aware even without the intervention of the brain at this early stage of his life. The brain is still empty with information at birth. The infant brain is not yet capable of storing, retrieving, and processing information. If you assume it does unknowingly, then it is still unknown.

Sensors, or exyzforms are the tools that create awareness. They are intuitive objects that come with mechanically embedded inscriptions. The word is derived from the Greek "exypnos" which means wise and "morfes zois" which means forms. Sensors are logically, mechanically, algorithmically, mathematically, structurally designed and inscripted. Intuitive sensors are the tale-tell signs of Awareness. The ability to inscript and configure without the help of the mind is known as Aneural Inscription.

Thus, AWARENESS, in its revised form, is the ability of an object to sense things through its sensors. Awareness emerges when an object is alive and equipped with sensors. This state of being is known as Sensoric Awareness.

— INTUITIVENESS
Intuitive or Logical State of Being

The third order of being is called Logical or Intuitiveness. In this state, the object is alive, aware, and intuitive. Intuitiveness is defined based on three determinants: (1) it uses energy, (2) it is driven by sensors, and (3) it makes choices. Intuitiveness emerges when aliveness, awareness, and logic are present. The sensorial self-ability to choose is the key factor that identifies when an object is logical or intuitive. Intuitiveness is the ability of an object to choose through sensory awareness.

The behavior to choose is dictated by the Law of the Second Option or the Rule of This or That. The Law is a Two Options Rule much like a flowchart that always stems with two options in every node. When information flows in the system, like in a conglomeration of switches, the signal produces a process called Binary LOGIC. The logic either turns on or off a system or subsystems that apparently becomes intuitive.

One common daily activity that involves the second option rule is walking. Walking always follows the law of second option. Our feet are designed to move in only two options every time. In slow motions, the foot can move backward, forward, sideways, upward, hop, kick, stride, or run. Whatever the preferences, the foot always has two and only two options at a given slice of time.

Like our feet, everything else always comes with only two options. These two options can be translated either 0 or 1, off or on, true or false, up or down, left or right, or maybe yes or no. Whatever the case, this dual option is called Inscriptional Logic or simply Logic.

When a baby prefers milk over porridge, the baby is intuitive. Although the baby may not know the nature of both foods, just by his sensorial experiences, his ability to choose mechanically shows intuitiveness. The baby's tongue is logical because it can differentiate good or bad food. The tongue is the sensor that makes the baby intuitive.

Plants are also logical because they are built with sensors. Their sensors are also governed by the two options rules. The baby and the plants may both be alive, aware, and intuitive, but they are not living and with life yet.

Intuitiveness is the ability to make logical decisions embedded without mental intervention. When such mechanical logics transform into Sequential Instructions, it gives rise to Logical Experiences. Sequential instructions produce rational experiences. The emergence shapes a system of wiseness known as Logical Intuitiveness. This rational state of being is known as Logical or Intuitive State of Being.

— ANEURAL CONSCIOUSNESS

The Aneural or Conscious State of Being:

The fourth order of being is Aneural Consciousness. In this state, the machine is alive, aware, intuitive, and conscious. To exhibit consciousness, an object needs four requirements: (1) it uses energy, (2) it comes with exyzforms, (3) it makes decisions, and (4) it can associate or match things. The ability to match is the crucial indicator of consciousness. Consciousness is the ability of an object to associate things with other things.

Consciousness does not need the presence of the brain. Plants are conscious not because they have brains but because they can codify. They can mechanically differentiate parameters like hot and cold, dry and wet, or day and night. Other creatures like corals, microbes, and micro-organisms can codify as well that evidently suggest they are conscious beings. Their actions to match or associate things without mental intervention is known as Associative Aneural Consciousness. Consciousness emerges because of the presence of energy, sensors, and codex.

Codexation or codification is the self-ability of an organism to match unknowingly things with other things. It is the key factor that identifies when an object is conscious or not. Babies at an early stage learn to play with colorful toys due to curiosity. Because of curiosity, they unknowingly learn things from taste, pain, sounds, and behaviors. When the food is bad, they dislike it. When the food is good, they eat it. When they are in pain, they cry. When not, they smile. Their sensoric actions of differentiating good or bad, without knowing the nature of pain, taste or sounds, is a form of associative consciousness.

In these cases, our babies, plants, and animals are alive because they consume energy; they are aware because they are equipped with sensors; they are intuitive because they can choose; and they are conscious because they can match or codify things. Since these objects move mechanically, interact sensorically, decide selectively, and associate things objectively, they are therefore Alive, Aware, Logical, and Conscious. This state of being is called Aneural Consciousness.

— INLEARNESS

Inlearness or Inform State of Being:

The fifth order of being is called Inlearness or Informed. In this state, the object is alive, aware, conscious, logical, and informed. The object is inlearned, if it has these five requirements: (1) it uses energy, (2) it responds to stimuli with its sensors, (3) it selects options, (4) it matches objects, and (5) it has information. It learns things without mental reasoning but through the aneural methods illustrated in chapter 4. The ability to have information is the crucial indicator of inlearness. Inlearness is the ability of an object to have information.

For example, when a sensor senses heat, its action is a kind of awareness. If it reacts to hotness or coldness, it is conscious and logical because it can differentiate things by design and choices. If it uses energy, it is alive. And, it is inlearned because its complex shape or form is inscriptionally designed with information. More sensors, more information, more informed. How is this so?

Remember that a heat sensor only triggers ON when it senses a temperature of hotness. When it senses coldness, it stays OFF or inactive. The heat sensor unknowingly but logically senses the difference between hotness and coldness. The sensor somewhat has a built-in sense of knowing much like our ears hear sounds without us *knowing* it mentally. The sensor experiences inlearness on its own due to its mechanical structure and embedded inscriptions much like an intuitive aneural network that stores information without the need of the brain. This non-mental behavior of possessing and processing information provided by inscription by design is called Aneural Inlearness.

The inability of the sensor to know it senses heat is the marginal line that separates life from non-life, from living to non-living, or from biozoic to abiozoic. Having the ability to know without knowing doesn't mean that a physical brain is required.

Knowing is simply a set of internal instructions gained from external sources like one's surroundings or from a complex combinations of sensors with their individual internal inscriptions. The ability to know via a set of instructions is part of inlearning.

Inlearning is the ability to possess and process information. These pieces of information are transformed into instructions. The self-ability to use information is the key that identifies when an object is inlearned or informed. An object is inlearned when it uses energy, driven by sensors, matches objects, selects options, and possesses information.

Information can be stored either naturally by sensoric designs or by the brain, or physically by programmers. Humans have "neurons" while machines have "aneurons". Neurons are parts of the human brain that store information, while aneurons are intuitive sensors that store information.

An IC, or integrated circuit, is a good example of aneurons. It can be programmed. It can be used on a machine to make decisions on its own as it logically analyzes and executes the data it gathers. The actual pieces of data the machine collects and executes are calculated by the machine itself through the inputs it receives from its sensors as it interacts with its environment. The data analysis and decision making don't come anymore from its programmer. The pre-programmed code and its physical design are the elements that create its actions and behaviors. The machine is "internally" aware of its surroundings because it behaves with no intervention from the programmer. If various sensors are attached to the machine and the appropriate prescribed signal is received, the machine can also be activated. The machine, in this state, is mechanically alive, sensorically aware, and aneurally inlearn.

Inlearness emerges when aliveness, awareness, consciousness, logic, and information are all present. This state of being is called Inlearned or Informed State of Being.

— LIVING

Livingness or Symbiotic State of Being

The six order of being is known as Living. Living emerges when aliveness, awareness, logic, information, codexation, and symbiosis, the six basic signatures of a living object, are present. The object (1) uses energy, (2) responds to stimuli with its sensors, (3) matches objects, (4) chooses options, (5) uses information, and (6) controls information. The ability of an object to use and control information is the crucial measure of living. The object uses the information to live symbiotically alone and side by side with other things.

— LIFE

Lifeness or Emergence State of Being

Lifeness or having life is basically defined as the formation or emergence of self. For Life to emerge and Self-realization to unfold, seven fundamental elements must be present, namely: (1) Aliveness, (2) Awareness, (3) Consciousness, (4) Logical, (5) Inlearn, (6) Living, and (7) Life. These whole seven measures are the new primary hallmarks or signature of a life-form that is alive, living, and with life. In short, Life emerges when one is alive and living.

The Emergence of Self is the telltale sign that indicates an object or entity is alive, living, and with life. Self-realization or selfness, on the other hand, emerges when aliveness, awareness, logic, information, codexation, symbiosis, and the emergence to recognize oneself are present.

One way to understand the seven orders of life is to compare these stages to someone with dementia. When a patient no longer uses information, although the said patient is still alive and with life, such person is no longer living. When the same individual can no longer take care of himself, make choices, and sense, the patient is still alive but no longer living and with life. And once the same patient stops eating (consumption of energy), said person is no longer living, with life, and alive.

We can also understand the seven orders by using inanimate objects. A piece of metal strip is a good example of this object. It reacts to both coldness or hotness. When it detects both temperatures, it acts as a sensor. Since it is a sensor, then awareness is present. Since it can expand or contract, it has inherent information. It is inlearned. Since it has options to choose from, its behaviors are logical, intuitive, and conscious. However, although it can store and use information embedded in its simple design, it is not living because it does not self-energize and therefore it is not alive.

Another way to understand the seven orders is through the next example using a pictorial circuit diagram:

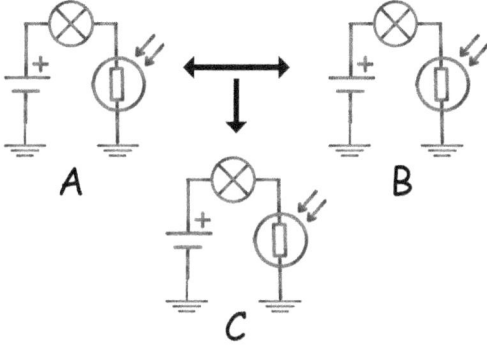

The electrical system is made up of a battery, a light bulb, and a sensor. In diagram A, the system is alive, aware, conscious, intuitive, and informed. When diagram A interacts with diagram B and diagram B interacts with diagram B, the system is considered living. If diagram A and diagram B produce diagram C, or diagram A creates diagram C, diagram A is not only alive and living, it has life now. Diagram A becomes a lifeform. Notice that the sensors in the circuit, the outside source the battery, and their individual inherent instructions are the elements that create life.

Thus, to redefine a life-form or an object with life, the entity is an intuitive object with embedded inscriptions that uses energy, perceives with sensors, codifies things, logical, inlearn, informs, and creates its likeness.

Overall, here are the key concepts on the states, orders, or degrees of becoming a being:

(1) Aliveness - the ability to consume energy provided by an external source (energy).

(2) Automated - the ability to do certain tasks based on a list of instructions (programs).

(3) Awareness - the ability to interact sensorically with its surroundings (intuitive sensors).

(4) Actuated - the ability to act on itself via external parameters (remote controlled).

(5) Animated - the mechanical ability to inscript itself (embedded inscription on structures).

(6) Aneural - the ability to codify, match, or pair things (associative consciousness).

(7) Logical - the ability to decide based on the law of the second option (intuitiveness).

(8) Inlearness - the ability to have information (informed).

(9) Living - the ability to control or use information (symbiotic experience).

(10) Selfness - the ability to realize oneself (self-emergence).

To have life, one must be alive and living.

CHAPTER 9 — IN VIVO MACHINA

"Everything exists because other things cause it to exist."

The expression might seem oxymoron, but according to the Theory of Generated Interim Emergence, everything exists because other things cause it to exist, otherwise, nothing ever exists at all.

For example, fire only exists because of these four basic essential elements: oxygen, fuel, heat, and chemical reactions. If one of these components is missing, fire will never ever emerge. This is like saying if a planet doesn't have oxygen and there are people-like aliens on that planet, these inhabitants will never know fire exists at all.

This is the same with music. A musical percussion, known as a xylophone, is made up of a set of colored bars of different lengths. In this instrument, music is not found anywhere. It doesn't exist in the xylophone. But when the bars are hit by a mallet (an outside force), sound or music is produced. Music is created because musical instruments like the xylophone cause it to exist. Without the right instrument, music will never be known.

Awareness does likewise. Awareness only exists when sensors are present. When an object is riddled with sensors and these sensors are triggered by an outside force, much like the job of the mallet above, the sense of awareness emerges. Without sensors, sensing or perceiving the surroundings will never occur, and the world around will never ever be experienced.

Same with ideas. Ideas exist because Mother Nature is equipped with information in forms of objects. Airplanes are invented because birds give us the idea of flying. Submarines are built because fish give us the idea of swimming. Houses are built because caves provide us with the idea of shelter. When fire was harnessed, infractures and tools not inherently found on earth were conceptualized and realized. Without Mother Nature and her creations, ideas and information will never exist at all.

Fire, music, awareness, information, and ideas are just a few examples of things that exist because other things cause them to exist. Consciousness, thoughts, emotions, dreams, animations, colors, and sensations are other examples of generated emergence also known as Interims. They are merely provisional by-products latent in essence. They are by-materials that can be sensed or noticed as if they are there, but in reality, they are simply the outset of emergence. This whole creation phenomenon is known as the Theory of Generated Interim Emergence (GenIE).

Life is also an interim. It emerges through GenIE. It evolves from being alive, living, and finally with life. The emergence is also governed by the same evolutionary seven orders of being:
1. Aliveness = self-consumes energy without neural reasoning.
2. Awareness = uses sensors without neural reasoning.
3. Intuitiveness = chooses this or that without neural reasoning.
4. Consciousness = matches things with things without the need of neural reasoning.
5. Inlearness = acquires and uses information using intuitive aneural memory networks.
6. Livingness = controls decisions and experiences symbiosis.
7. Selfness = self-formation and self-realization.

Engineered lifeforms can also be created by employing the same evolutionary orders of life through the following stages:
stage 1 - mechanical animation energized by external source
stage 2 - sensoric awareness triggered by sensors

stage 3 - codified consciousness using sensors to copy

stage 4 - algorithmic logic using sensors to choose

stage 5 - informed inlearning using sensors for information

stage 6 - symbiotic living using sensors to socialize

stage 7 - emergence of self by design and information..

Through these stages, the emergence of a mechanical lifeform, an in vivo machine, can be accomplished by building it from scratch and gradually making it a living thinking machine. A more sophisticated version of our previous biomechatronic spider will be used as our next prototype. The machine is not only made up of a conglomerate of axles, gears, beams, motor, and battery but also mixed with various sensors, programmable microcontrollers, and rechargeable power supply.

Basically, the machine can stride either forward or backward when an outside force (my hand) is applied on it. The applied force is the push or pull that sends a signal to its legs and parts to set it in motion. When the applied force stops, the machine stops. The machine can only be in motion through this outside force.

The force, called the cue, is provided by the battery. The power supply is not part of the machine but an external system that can be attached, internally or externally, to the object. Cues are signals that switch on the set of inherent instructions or embedded procedures owned by the structural design of the spider. Cues can be mechanical, electronic, hydraulic, sound, light, electrical, magnetic, digital, or anything that can turn on or off its whole system. The force and source of energy are the essential factors that make the machine in motion.

— MECHANICAL ALIVENESS

The force is the switch that triggers simultaneously all the various parts, sub-systems, and instructions on the machine. Every part of the machine is an intuitive object with embedded instruction. Its electronic eyes, ears, and various sensors, down to its long bi-segmented legs are all intuitive objects. All these parts are pieces of instructions that form as a single program of the whole system. When triggered, the individual instructions of each part are simultaneously activated. When all the parts are continuously supplied with force, the list of instructions is activated simultaneously and its whole system is animated automatically, generating the phenomenon known as Animation by Design (Lawsin, 1988).

But, where does the machine get its constant supply of force to animate it continuously?

The machine gets its supply of force by "consuming food" from external sources. The food or energy can come from batteries, sunlight, chemical, mechanical, sound, electrical, motion, or thermal. The machine, in this case, can harvest its energy either by docking on a charging station or powering itself via solar panels. Once its energy supply is fully charged, all its mechanical parts and instructions are initiated simultaneously and synchronized subsequently making the machine animated one more time just like a one big organism. Take note that the machine is constantly activated by the force or cue that powers the machine and not by the energy. Force only exists when energy is present.

When the machine self-consumes its energy in the charging station, aliveness or being alive emerges, and the overall system and subsystems of the machine gradually becomes automated, animated, and alive.

When does the machine know it needs juice or when its stomach is empty so to speak?

— **SENSORIC AWARENESS**

A predesigned sensor, for example, called a voltage indicator can be wired on the machine. It is an electronic device designed to detect when a battery is low or high. When it detects a low level, the sensor sends a signal to a receiver, which is another sensor designated, for example, to activate a light or a motor that eventually alerts the machine to go to a charging dock station and recharge itself.

The machine, in this stage, is considered alive and aware because it self-consumes energy and uses its intuitive or wise sensor to interact with its environment. When a cue is detected by its mechanically predesigned or naturally evolved codified sensor, awareness emerges and its whole system is activated and thus every instruction in every part of the machine is processed. The machine behaves like a spider because its physical design resembles the anatomy of a spider. Thus, it crawls.

How does the machine know when to respond to a cue?

— INTUITIVE LOGIC

The machine responds to the cues through its codified sensors. A sensor always comes with two options. This is due to the Law of Second Option, the This or That Rule, or the Flowchart Effect. Everything is generally governed by two choices. And if there is a third option, the third option becomes the second option.

Our ears, eyes, and nose are special sensors in their own ways. The ears can only hear but can't smell or see. The eyes can only see but can't smell or hear. The nose can only smell but can't see or hear. Each one of these sensors is unique. The signals they receive are also unique because the ears only receive sound waves, the eyes receive light waves, and the nose receives olfactory waves. Because of these sensory uniqueness, the physical design of a codified sensor and the correct signal the sensor receives are directly proportional to each other. Sensors are logically intuitive due to the Law of Second Option.

When the machine receives a cue or signal, the signal activates the various logic found on its sensors, parts, and systems. The logic is then translated into numerous rationalities as it flows into its whole structures from simplicity to complexity much like a flowchart that runs a sequence of branches, codes, flows, functions, or instructions from top to bottom, left to right, corner to corner. When the signal enters this flow of binary logic, the machine becomes intuitive.

The physical design of the sensor makes the logical decisions to either accept or not to receive the signal. If the right signal enters the right sensor, the system is then activated. If not, it will stay off. Notice that a brain is not necessary here to process information. What it takes is the correct signals and the right design of the codified sensors to be logical.

A mechanical gizmo can be used as a substitute for any sensor. For example, a device made up of legos can be built to sense its environs through touching, hearing, smelling, tasting, seeing, or balancing, A protruding wooden stick can be assembled to receive and send a signal when an object touches a wall and reverses itself to avoid the obstacle. A contraption can also be used to activate something like for example a light that is refracted on the surface of a magnifying mirror and converged inside a radiometer to spin its vanes.

The sensors in our machine are like these contraptions. It is turned on when the right type of signal is applied or received. Signals can be anything. They can be light, water, air, sound, pressure, electricity, weights, temperature, or even gravity. Signals and sensors are mutually designed with inherent individual specific tasks. A mirror is designed to receive or send light. A tuning fork is designed specifically to receive or send sound waves. A wooden stick is designed to receive or send touch pressure. Whatever the case is, the structural design of a sensor is extremely important in acquiring the correct signal. Every sensor has its own distinct matching signal.

How does the machine become conscious?

— CODIFIED CONSCIOUSNESS

Consciousness has been defined in various ways. Many believe that consciousness emerges from the brain. Some argue that the basis of consciousness lies in the nervous system. Some even suggest it comes from transcendental or mystical experiences. Others claim that consciousness is the soul or spirit that powers the body. Of course, these beliefs are clearly unfounded. Why?

According to the theory of Aneural Consciousness, consciousness is an interim that does not need to emanate from the brain due to the following facts where in many domains there are organisms that are:

(i) alive but without brains,

(ii) alive but not conscious,

(iii) alive but not aware,

(iv) alive but not living,

(v) aware but not conscious,

(vi) aware but without brains,

(vii) conscious but not aware,

(viii) conscious but not self-conscious.

(ix) conscious but without brains,

Furthermore, if we recall our discussions on the Codexation Dilemma, consciousness is codified. This means that if an entity or object can associate its thoughts or needs with something else, then this object or entity is conscious. The something can be an object, a word, a label, a tag, a name, a definition, a description, a symbol, a behavior, or even a gesture.

For example, birds can avoid predators, build nests, and feed their babies. They can associate avoid with predators, build with nests, and feed with babies. These behaviors tell us that they are conscious. We called this measure of comparative matching, labeling, tagging, naming, or one-to-one correspondence as Associative Consciousness (AC).

Aside from AC, we also came out with the abridge definitions of consciousness:
1. Consciousness is a correlation or association.
2. Consciousness comes with collaborative determinants.
3. Consciousness is inscripted by design.
4. Consciousness is codified.
5. Consciousness is a generated interim.
6. Consciousness is materialized by information.
7. If I can match x with y, therefore I am conscious.

From these findings, we concluded that an object is conscious if it can match, label, or associate things with other things.

When a sensor associates a low battery with a charging station, it shows consciousness. When a machine crunches data by itself, it provides clues that somehow it "knows" where to find the charging station by computing its exact distance. The exact data that triggers the machine to locate the station is not coming from me, the programmer. The hidden generated calculation is a form of a latent consciousness that comes only from the machine through its scores of transistors which technically are sensors. This type of behavior is called mechanical consciousness.

The machine is also conscious, when it "knows" how to avoid obstacles like when it detects a wall and reverses itself for a safer new ground. It can associate obstacles with danger, avoid with reverse, and safe with new roads. The machine, in this case, is also conscious because it can codify through its intuitive sensors.

Consciousness is the demarcation line that separates the living from the non-living things. It transpires through codified sensors or by coding. It emerges when one is alive, aware, intuitive, and conscious.

How does the machine acquire information?

— INFORMED INLEARNING

Traditionally, information is acquired in two ways: by choice or by chance. In humans and animals, it does not automatically exist in the brain. It follows a process. It is codified, stored, retrieved, and processed in the brain. The acquisition of information is a step-by-step series of instructions. It is sensed, perceived, interpreted, identified, and gained. It is acquired by touching, seeing, hearing, tasting, thinking, or smelling. It is an individual attribute 100% universally congenital in living things.

In autognorics, information is an inherent property owned by everything. When information is inherently acquired gradually one by one, bit by bit, and accumulated together piece by piece, jumping in a queue, the linkage of information emerges as instruction. This inherent emergence of instruction by design is known as Inscription.

A machine can store information through inscription by design or by coding. By design, the information is already stored in its structural design and reused over and over much like a whistle that always produces the same pitch of sound when blown. The information of the sound is stored in the whistle through its structural design. The blown air is the trigger that activates and runs the inscription of the sound in the whistle. Through its design, the whistle activates a series of internal information without neural reasoning, another example of a brain without the brain.

By coding, machines store information using microprocessors. Through integrated circuits, machines can be encoded like the traditional ways humans acquire, store, codify, retrieve, and process information. However, one of the caveats in making a functional thinking machine is that a huge amount of information needs to be stored. Thus, a bigger storage, a larger energy source, and a complete structural design must be undertaken to accommodate this dilemma.

The pieces of information below can also be used as a guide in shaping a thinking living machine.
1. Humans acquire information from their environments.
2. Information is acquired by choice or by chance.

3. Nature is the original supplier of Information.

4. Nature is the very first brain made up of stored information, a database of objects.

5. Humans copy or mimic things from the environment.

6. The environment makes you or programs yourself.

7. Thinking doesn't prove existence and consciousness.

8. The brain of a baby is empty with information at birth.

9. Information flows from nature into the brain.

10. Information physically comes from outside sources and goes abstract inside the mind.

11. No humans can think of something without matching their thoughts with something physical.

12. Sequential instructions produce rational experiences.

13. Information turns into instructions and procedures.

14. Everything is an intuitive object with embedded instruction.

15. Everything has a beginning and emerges due to other things.

Since, our machine now is alive, aware, intuitive, conscious, and inlearn, how does it become a living thing?

— SYMBIOTIC LIVING

Typically, it is thought that when an object gains, stores, codifies, retrieves, and processes information, the object is perceived as an intelligent entity. Humans, although notorious as short-term thinkers, are believed to be intelligent because they have brains. Some man-made objects like computers and robots are artificially intelligent as well because they can process information even if they have no brains. Bacteria, algae, fungi, plants, animals, and other natural objects like rocks, air, water are constantly experiencing some of these functions too due to inscription by design and not because they have brains. These non-living objects. like everything else, are embedded with information. They can store, codify, retrieve, or process information through their shapes and designs.

When an object uses and controls information, makes social interactions, and experiences things around, it is alive, aware, intuitive, conscious, informed, and living. When it lives and interacts side by side with other things, it is a living object.

When a machine sends a signal to an object like an intuitive wall, and the wall receives the signal and responds back by flashing a light, and the machine accepts the signal by crawling forward and backward suggesting it receives the cues, the whole interactions of the machine and the wall show mutually a simple display of livingness. Both objects mutually use information, make internal decisions, and interact with each other.

Ergo, to make our machine living, six elements must be present, namely: aliveness, awareness, intuitiveness, inlearness, consciousness, and mutual interactions.

How does the machine become a lifeform and emerge as SELF?

— SELF EMERGENCE

Peanut is one of the experimental dogs that I've used in my research on the Caveman in the box and Bowlingual experiments. He is a chihuahua who happens to live once in a chaotic environment. His experiences in his previous home drastically traumatized his life. This trauma of nervousness and shaking manifests whenever people are around. He stays in the corner most of the time just to avoid everything. His head and body almost touch the ground when he walks around. He is also scared of loud noises. His left eye is battered, and he squeals when someone carries him.

When we adopted him, his tormenting behaviors did not go right away. Trust and love were not in his vocabularies. This is where we eventually introduced Zero to him. Over time, Zero and Peanut bonded together and Zero became his big brother. Many things have changed in his life when Zero became part of his life. He became friendly, naughty, cheerful, playful, and barks a lot nowadays. He learned Zero's behaviors and skills and transferred some of his skills as well to Zero.

As both get old, Peanut suffers blindness. In the onset of his blindness, he always bumps his head on the walls wherever he goes in the yard and the garage. Although he has a picture of the size of the place when he was not blind yet, the whole floor plan is still clear in his mind. Before, he gets confused walking from the backyard to the garage. Now, he can freely navigate himself from one place to another by avoiding the obstacles by merely walking backwards a few inches while cautiously feeling the correct way to safety.

He also continuously communicates with us everyday with four types of sound-signals. When he wimps, it means he wants to go out to pee or poo. When he barks, it means he wants to get inside the house. When he yelps, it means he is excited. When he howls, a behavior he learns from Zero, it means he is lonely, needs something, or in pain.

Just like his thoughts, feelings, behaviors, and formation, self-realization emerges internally in him because of information. Thoughts, feelings, behaviors, formation, and self-realization are recreated by the successions of information we gained over time from our surroundings. They are like modular programs in a flowchart that are activated selectively anytime when the right cue is received. When we smile, we feel happy because there is a series of instructions attached with smile and a series of instructions that makes us happy. When we think of death, we think of sadness because there is a series of instructions connected with death and a series of instructions that we attach with sadness. Thoughts, feelings, behaviors, formation, and self-realization emerge because a unique sequence of information or instructions cause each one of them to exist.

When the right conditions, right elements, right designs, and right signals are present, the right yield emerges. Design emerges as a sequence of code and as a tool to process information. Rational experiences emerge because of intuitive sequential instructions. The emergence of self unfolds when the Law of the Seven Inscription transpires:

1. It can transit energy into its design, the key factor of life.
2. It can transcribe design to information, information by design.
3. It can transfer information into actions or movements, the first universal instruction.

4. It can transmit motions into mechanical, repetitive, or autonomous actions, the animation effect.
5. It can translate the mechanical persona (feeling, thinking, behaving) into an inlearned experience.
6. It transcends the experiences into the formation of self.
7. The formation transforms into the emergence of Self.

CHAPTER 10   **INSCRIPTIONAL PHYSICS**

"Inherently, everything owns a series of instructions."

Inscriptional Physics is a discipline that deals with the natural equations or set of instructions embedded in every object, from the smallest particle to the massive matter in the universe. Its main goal is to find the mother of all equations, the grand inscription, that governs everything. It also seeks every unique inherent equation owned by every object, from nonliving to living things, and everything in between.

Inscriptional Physics also tackles the existence of the non-existence of numbers, the Codexation Dilemma, one of its many paradoxes. It also tries to solve many math problems like measurement by division, reciprocal of zero, and division by itself by introducing new methods such as the rules of zeroes, the axiom of x, orders of mathematical operation..

One of them is the units of measurement. Like numbers, units also follow the same four basic operations of mathematics: such as addition, subtraction, division, and multiplication. There are also other mathematical rules that need to be reconsidered. One of them is the rules of zeroes.

The Rules of Zeroes state that::

(a) $0 + 0 = (1+1)*(0) = 2 * 0 = 0$
(b) $0 - 0 = (1-1)*(0) = 0*0 = 0$
(c) $0 * 0 = 0^1 * 0^1 = 0\,^(1+1) = 0^2 = 0$
(d) $0 / 0 = 1*0/1*0 = 1/1*0/0 = 1*0 = 0$

Since the proofs are evidently valid without contradictions, we can apply these rules also with variables such as the letter X. Let us call these rules as the Axioms of X.

(a) $x + x = 2x$.
(b) $x - x = 0$.
(c) $x * x = x^2$
(d) $x / x = x$

We were told that a variable, number, expression, or object, when divided by itself, is always equal to one. This rule seems valid. But if you examine this rule once again, there is actually a glitch. Here is an example:

(a) 6 kg + 2 kg = 8 kg.
(b) 6 C - 2 C = 4 C.
(c) 6 m * 2 m = 12 m.

(d) 6 s / 2 s = 3 s/s

When we follow the rule x/x=1, we notice in our examples that the fourth statement is obviously wrong. Commonly, people think that the answer is 3 seconds. But the result should be stated correctly as 3 sec/sec or simply 3 without a unit. But because of the rule x/x=1, the right answer should be 3 and no unit of measure (unitless) at all. However, to preserve the prescribed correct unit of the equation, and to correct this glitch, I introduced the superscript degrees symbol (^0), since anything raised to the power of zero is always equal to one. This is represented as $X^0=1$.

Thus, in our example, instead of using s/s, we can preserve the unit second(s) by raising the unit to the power of zero. In this case, the s/s can now be notated as $(s/s)^0$ or s^0. The s^0 preserves the integrity of the unit, which is the unit second. And since s/s = 1, and $s^0=1$ as well, then the result of the equation is not disturbed at all.

Therefore, to rewrite the fourth equation with the correct unit:

- 6 s / 2 s = (6/2)(s/s) = (6/2)(s^0) = $3s^0$ (3*1 = 3).

The unit can now be both visible and invisible depending on one's preferences.

However, rationally and mathematically, this division by itself can be co completely altered based on the rules of zeroes and the axioms of x. If we follow them, the fourth equation's unit will no longer be unitless since the unit second(s) is now visible.

Thus, to rewrite the rule on division by itself; "When a variable, number, expression, or object is divided by itself, the result is always equal to itself.".

Henceforth, when we divide negative by negative, the answer is negative. For example, -1/-1=-1.

- -1 / -1 = -1
- -1 = -1 * -1
- -1 = (-1*1) * (-1*1)
- -1 = -1(1*1)
- -1 = -1(1^2)
- -1 = -1

Also, following the rules of zeroes, we can now have a multiplicative inverse for zero. When we divide 0/0, it can also be expanded to (0/1)*(1/0). Notice that 1/0 is the reciprocal of zero. Since 0 has now a reciprocal, then (0)*(1/0) or 0*1/1*0 = 0/0 = 0.

Thus, in conclusion, if we accept 0/0=0, then we can now rationally and mathematically correct three major problems in math: namely, the unit of measurement by division, reciprocal of zero, and division by itself.

Now, let us have a counterexample on these said rules.

- 10/10 = 10
- 10 = 10 * 10 = $(10)^2$ = $(10)(1^2)$ = 10 * 1 = 10
- therefore: 10/10 = 10

This math diversion is what I called the Red Herring Trick.

Now, we were told that if there are 6 slices of pizza and there are 2 kids in the room, how many pizzas will each kid get? And if there are 10 pencils on a table and 10 students in a class, how many pencils will each student have?

Using the Red Herring Trick, the answer is of course one(1) !

- 10/10 = 1/1 = 1

Another math issue that needs to be corrected is the Mathematical Orders of Operations known as PEMDAS or BODMAS. These orders must be revised due to the five collaborating pieces of evidence that I have deduced while working on the mathematical side of my single theory. They are:

Proof #1. The operators, plus (+), minus(-), cross(x), obelus(/), and exponent(^e) can produce answers or results. They can add, subtract, multiply, divide, and raise numbers, respectively. When engaged mathematically, they provide some fixed answers. However, a parenthesis cannot produce an answer or result, at all. It acts only as a simple holder of numbers inside its brackets.

For example, 4 + 5 = 9, 9 - 8 = 1, 8 / 4 = 2, 7 x 3 = 21, and 6 ^2 = 36. All the operands on these notations or expressions produce some kind of result or answer. But when we put a number inside a bracket, like (4), the parenthesis is not doing anything except that it is just embracing the number inside its cell. It does not or cannot produce an outcome.

Now, if we consider the parenthesis as a property of identity, where it carries an exponent, except one, and becomes a part of the exponential family, then the parenthesis can now produce a result. For example, when we raise it to the power of zero [(4)^0], it produces a result equivalent to one [(4)^0=1]. Here, PEMDAS can now be re-arranged logically as E M D A S, without the parenthesis.

Proof #2: MDAS is the forerunner of all mathematical operations. They are the first four operands we learned at an early age in our lives. They are the first four original operators we loved to play with numbers when we were studying math. From them we progress to exponents and factorials. Due to this reason, learning from simple to more complex operations, we can place MDAS first in the order, and add the E last, like in MDASE.

Proof #3: Aside from the first magnificent four, plus the exponent, another essential operand must be also added to the orders of mathematical operation. This is the factorial operand represented by the exclamation point (!). Factorial also generates a result, for example, zero factorial is equal to one (0!=1). Now, if F is included in the series, the complete order of operations becomes MDASFE.

Proof #4: Scientifically, when we solve for X in a non-numbers-all-variables equations, we usually solve them by simplifying the equation first. Once simplified, we solve the value of X.

Proof #5: The term equation stalemate, or equation freeze, is a term coined to describe an equation that cannot be solved without deviating from the rules of operations. Due to this technique mathematical problems can now be conquered in different ways through critical thinking, intuitive imaginations, and unexpected deviations..

From all these rational, mathematical, psychological, scientific, and interpretational pieces of evidence, MDASFE should be used instead of PEMDAS. And for those mnemonic fans, MDASFE stands for *MDAS F*or *E*ver!

— THE ELEGANT EQUATION

The Theory of Generated Interim Emergence, also known as the single theory of everything, teaches us that for something to exist, it needs two things: materials & instructions. This statement when translated into equation states that "the summation of its external intuitive materials and its internal embedded inscriptions is equal to the emergence of an interim entity".

The Variables:

- GE = generated emergence
- IO = intuitive objects
- EI = embedded inscriptions
- n = number of materials
- i = number of instructions

- a = the interim entity

The Derivation:

- if: GE emerges due to both objects & inscriptions
- then: GE is equal to n(IO) and i(EI)
- or: $\sum GE = \sum n(IO) + \sum i(EI)$
- thus: $\sum I = \sum A + \sum N$
- where: Lw is the derived unit of measurement.

The Verbatim:

"The emergence of an interim *(I)* is equal to the number of its materials*(A)* plus the number of instructions on its inscriptions*(N)*." ~ the Single Theory of Everything.

— NEURO TOMOGRAMS

In this section, I also incorporated some of the slices or images of my neuro computed tomograms for future studies. I imagine that someday an expert in neuroscience would analyze these brain photographs for the sake of science. These negatives were supposed to be integrated in my first self-published book, Creation by Laws.

In the same book, I deliberately published my manuscript in a draft format. The purpose was to show to the public how my mind freely expresses my thoughts in real time without following any general rules. In fact, in real life, this is surprisingly how my brain works. It sounds chaotic, but ingeniously it led me to the orders of the Single Theory of Everything.

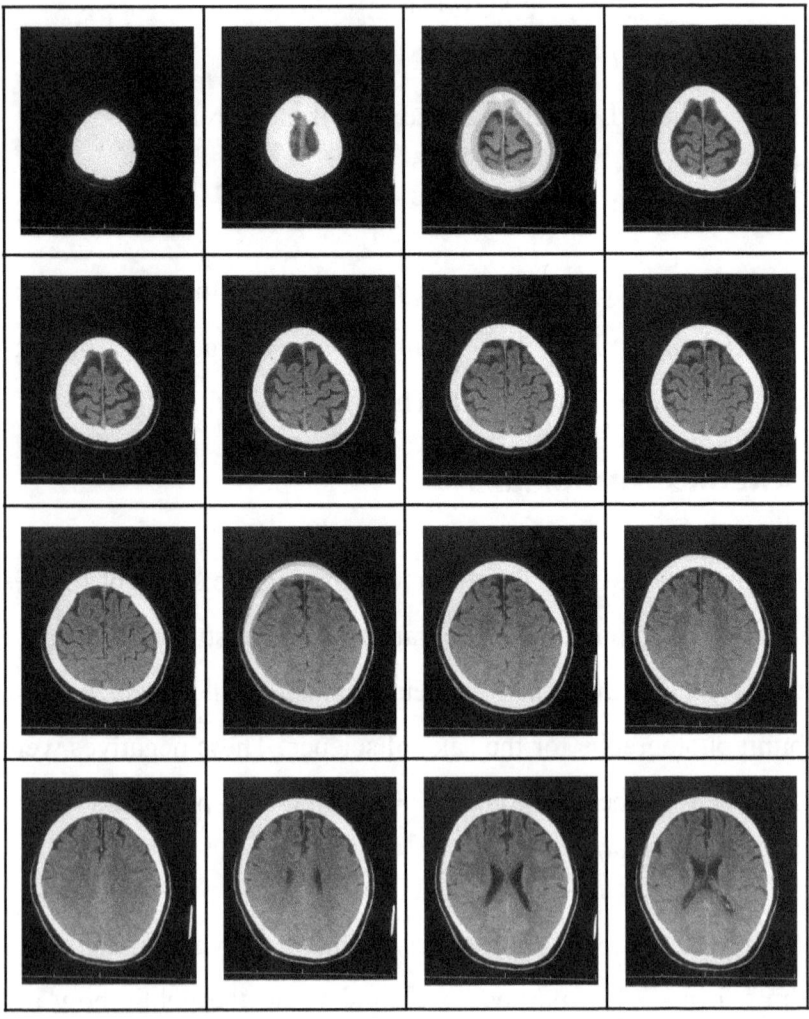

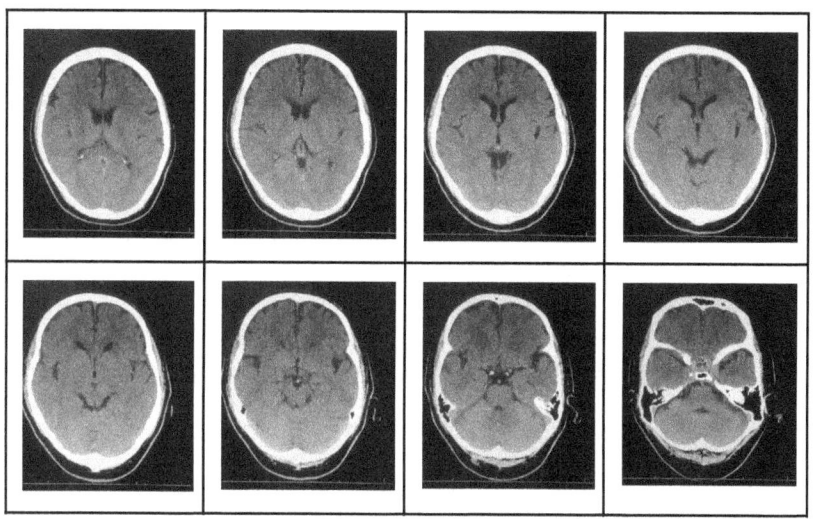
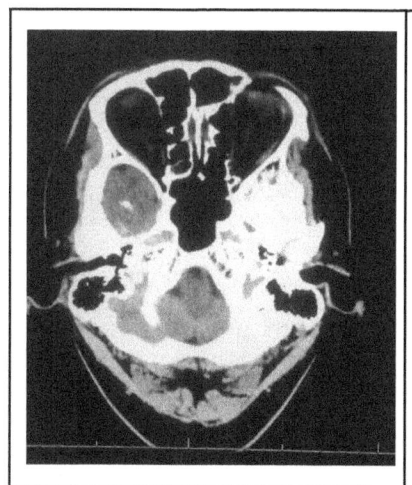
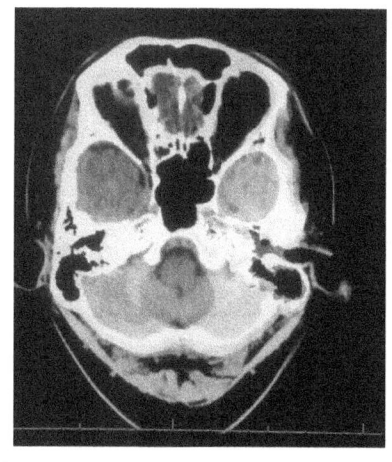

— GNOS

/gnōs/

noun An intuitive aneural network unit much like a neuron which can recreate or design itself based on its acquisition of information.

etymology Sanskrit *jna-* "know;" Avestan *zainti-* "knowledge," Old Persian *xšnasatiy* "he shall know;" Russian *znat* "to know;" Latin *gnoscere* "get to know," *nobilis* "known, famous, noble;" Greek *gignōskein* "to know," *gnōtos* "known," *gnōsis* "knowledge, inquiry;" Old Irish *gnath* "known;" German *kennen* "to know," Gothic *kanjan* "to make known."

γνῶσις Self-knowledge, introspection, inlearned, self-discovery, self-imagination, self-informed, a dreamer.

*Lawsinist An individual with a passion to hunt or seek the natural laws of the universe in order to unearth the central dogma of life known as the Single Theory of Everything.

Lawsinists are inspired to uncover equations in every object, formulas in every phenomenon, and the single theory in every nook and cranny in the cosmos. They always see beauty and elegance in numbers, equations, and the natural laws of the universe. They are sometimes called Inscriptionists.

ABOUT THE AUTHOR

JOEY LAWSIN, BSAE, PgDE, MBA, describes himself as a revisionist independent thinker who has a deep passion in solving the origin of information and the single theory of everything through investigating an unsuspecting subject known as Inscription by Design.

Most of Joey Lawsin's scientific, theological, philosophical, and technological seminal works are collected in his books.

Scientifically, his research interest is focused primarily on Inscription by Design(ID), a theory that led him to the discovery of the Single Theory of Everything. He asserts that everything like consciousness, life, and self are all nothing but illusions that seem to exist only because other things cause them to exist. The central controlling idea of his ID is threefold. First, the Codexation Dilemma. The theory that asserts abstract ideas cannot be transformed or iCodexated into physical realities without the external material inherent world. Second, Originemology. The study that postulates Nature is the brain, mother, and keeper of Information. And third, Generated Interim Emergence. It posits that everything is made up of empty space and invisible shape, the building blocks of life, which are intuitive objects with embedded instructions which when energized cause the animation phenomenon we humans call Reality.

Theologically, he beautifully developed various new radical schools of thought on creation, evolution, life, god, and the bible. Focusing on the hows and whys, he made important discoveries about why the bible is the key evidence that provides concrete proof there is no god, why life is more of natural inscription than divine creation, why reality is a mechanical illusion, how man created god, and how religious beliefs of the past is destroying humanity.

Philosophically, he advances old philosophical views mastered by well-known philosophers and scholars to new heights of ideological paradigms such as the Nature or Nurture maxim; I think, therefore, I am dictum; and the philosophy of the Mind Theory. His books also contain original *paraduoxical* quotations that were subsequently rooted from his experiments.

Technologically, in a quest to unravel the mystery of consciousness and to build a living machine that recreates thoughts, emotions, and behaviors, he formulated five new fields of study on robotics, namely: Neurotronics, Homotronics, Biotronics, Autognorics, and Dimetrix. His skills in programming and electronics are demonstrated in his various projects such as controlling robots or machines remotely using the internet, a personal website, via wifi, an ethernet, a TV remote control, a computer monitor, a keyboard, mouse, bluetooth, an iphone, ipad, and even voice commands. One of his bragging rights is that he programmed a search engine and a dos-driven menu before Google and Windows were conceptualized.

As a progressive humanist and revisionist, he actively promotes a cause that encourages humans to set aside their belief systems which are slowly destroying humanity; coexist and reorganize as one race in order to move forward quickly into the future of space exploration; and restore mother earth to her pristine abundance before modern civilization becomes history and information dies.

As a dedicated Autism advocate, he immensely desires to find solutions on how autism can be reverse engineered based on his seminal findings on Originemology; and how the damaged parts of the human body can be replaced using intuitive objects, embedded inscriptions, and generated emergence medicine.

Joey Lawsin is the author of several books including previous paperbacks, such as Creation by Laws (2008), Evolution of Creation (2010), Inscription by Design (2018), Originemology (2022), The Single Theory of Everything (2020); technical manuals such as The Biotronics Project (1988) and Autognorics (2022); and academic textbooks in Physics (ed. 1991, 1993, 1996).

www.ingramcontent.com/pod-product-compliance
Lightning Source LLC
Chambersburg PA
CBHW070229180526
45158CB00001BA/216